The Proper Ambition of Science

London Studies in the History of Philosophy is a unique series of tightly-focused edited collections. Bringing together the work of many scholars, some volumes will trace the history of formulations and treatment of a particular problem of philosophy from the Ancient Greeks to the present day, while others will provide an in-depth analysis of a period or tradition of thought. The series is produced in collaboration with the Philosophy Programme of the University of London School of Advanced Study.

Series Editors: Tim Crane, Thomas Pink, M. W. F. Stone and Jonathan Wolff.

1 **Humanism and Early Modern Philosophy**
Edited by Jill Kraye and M. W. F. Stone

2 The Proper Ambition of Science
Edited by M. W. F. Stone and Jonathan Wolff

London studies in the history of philosophy
Series editors: Tim Crane, Thomas Pink, M. W. F. Stone and Jonathan Wolff.

London Studies in the History of Philosophy is a unique series of tightly-focused edited collections. Bringing together the work of many scholars, some volumes will trace the history of formulations and treatment of a particular problem of philosophy from the Ancient Greeks to the present day, while others will provide an in-depth analysis of a period or tradition of thought. The series is produced in collaboration with the Philosophy Programme of the University of London School of Advanced Study.

1 **Humanism and Early Modern Philosophy**
Edited by Jill Kraye and M. W. F. Stone

2 **The Proper Ambition of Science**
Edited by M. W. F. Stone and Jonathan Wolff

The Proper Ambition of Science

Edited by M. W. F. Stone and Jonathan Wolff

London and New York

First published 2000 by Routledge
11 New Fetter Lane, London EC4P 4EE

Simultaneously published in the USA and Canada
by Routledge
29 West 35th Street, New York, NY 10001

Routledge is an imprint of the Taylor & Francis Group

Typeset in Times by Curran Publishing Services Ltd.
Printed and bound in Great Britain by
St Edmundsbury Press, Bury St Edmunds, Suffolk.

British Library Cataloguing in Publication Data
A catalogue record for this book is available
from the British Library

Library of Congress Cataloging-in-Publication Data
The proper ambition of science/edited by M. W. F. Stone and Jonathan Wolff.
240 pp. 13.8 x 21.6 cm (London studies in the history of philosophy: 2)
I. Science – Philosophy. II. Science and civilization. I. Stone, M. W. F. (Martin William Francis) 1965– . II. Wolff, Jonathan. III. Series.
Q175.3.P786 2000
501–dc21 99 39024
CIP

ISBN 0–415–18617–X

Contents

Contributors

Nancy Cartwright is Professor in the Department of Philosophy, Logic and Scientific Method at the London School of Economics (part-time), Professor of Philosophy at the University of California at San Diego, and Director of the LSE's Centre for Philosophy of Natural and Social Science. Her books include *How the Laws of Physics Lie* (1983), *Nature's Capacities and Their Measurement* (1989), *Otto Neurath: Philosophy Between Science and Politics,* with Thomas Uebel *et al.* (1996) and *The Dappled World: A Study of the Boundaries of Science* (1999).

Christopher Hookway is Professor of Philosophy at the University of Sheffield, having previously taught at the University of Birmingham. His publications include *Peirce* (1985), *Quine: Language, Experience and Reality* (1988), *Scepticism* (1990) and *Truth, Rationality and Pragmatism: Themes from Peirce* (2000). He was President of the Peirce Society for 1995.

J. R. Milton is a senior lecturer in the Department of Philosophy, King's College London. He is interested in late medieval and early modern science and philosophy, and is currently working on a biography of John Locke.

Dermot Moran is Professor of Philosophy at University College Dublin. A graduate of University College Dublin and Yale University, he has taught at Yale University, Connecticut College, Queen's University of Belfast, and Maynooth College. He is the author of *The Philosophy of John Scottus Eriugena: A Study of Idealism in the Middle Ages* (1989) and other studies in medieval and contemporary European philosophy. He is currently Editor of the *International Journal of Philosophical Studies*.

David Papineau is Professor of Philosophy of Science at King's College London, having previously taught at Cambridge and Reading

Universities in England and at Macquarie University in Australia. He is author of *For Science in the Social Sciences* (1978), *Theory and Meaning* (1979), *Reality and Representation* (1987) and *Philosophical Naturalism* (1993). He was Editor of the *British Journal for the Philosophy of Science* from 1993 to 1998.

Aaron Ridley is a Senior Lecturer in Philosophy and Associate Director of the Centre for Post-Analytic Philosophy at the University of Southampton. His books include *Music, Value and the Passions* (1995) and *Nietzsche's Conscience: Six Character Studies from the 'Genealogy'* (1998).

G. A. J. Rogers is Professor of the History of Philosophy at Keele University. He is the author of *Locke's Enlightenment* (1998) and the editor with the late Peter Nidditch of the *Drafts of Locke's Essay Concerning Human Understanding* (Clarendon Edition of the Works of John Locke). He has also edited eight other books and is the author of over sixty articles and parts of books. He is the founder-editor of the *British Journal for the History of Philosophy.*

M. W. F. Stone is Lecturer in the Philosophy of Religion, King's College London. His main research interests lie in the field of medieval philosophy. He is the author of the forthcoming two-volume work *The Subtle Arts of Casuistry* and has also edited with Jill Kraye *Humanism and Early Modern Philosophy* (2000), and published articles on medieval and early modern psychology and ethics.

R. W. Sharples is a Professor of Classics at University College London and current Head of the Department of Greek and Latin. He is the author of *Stoics, Epicureans and Sceptics* (1996) and has published on Plato, Theophrastus, Cicero and Alexander of Aphrodisias.

Thomas E. Uebel is a member of the Centre for Philosophy in the Department of Government at the University of Manchester. His research concerns epistemology, philosophy of social science and history of philosophy of science; his publications include *Overcoming Logical Positivism From Within* (1992) and (with N. Cartwright *et al.*) *Otto Neurath: Philosophy Between Science and Politics* (1996).

Jonathan Wolff is Reader in Philosophy at University College London. He was the Founding Director of the Philosophy Programme in the University of London School of Advanced Study. His publications include *Robert Nozick* (1991), *An Introduction to Political Philosophy* (1996) and papers in contemporary political philosophy and the history of political thought. He is Editor of the *Proceedings of the Aristotelian Society.*

1 Introduction

Jonathan Wolff

> We make our beginning with a change which set in at the turn of the past century in the general evaluation of the sciences. It concerns not the scientific character of the sciences but rather what they, or what science in general, had meant and could mean for human existence. The exclusiveness with which the total world-view of modern man, in the second half of the nineteenth century, let itself be determined by the positive sciences and be blinded by the 'prosperity' they produced, meant an indifferent turning-away from the questions which are decisive for a genuine humanity. . . . It excluded in principle precisely the questions which man, given over in our unhappy times to the most portentous upheavals, finds the most burning: questions of the meaning or meaninglessness of the whole of this human existence. . . . Scientific, objective truth is exclusively a matter of establishing what the world, the physical as well as the spiritual world, is in fact. But can the world, and human existence in it, truthfully have a meaning if the sciences recognize as true only what is objectively established in this fashion . . .?
>
> (Husserl 1970: 5–7)

In the Preface to his *The Crisis of European Sciences*, Husserl presents a picture in which science, having banished the most burning of human questions from its domain, comes increasingly to dominate culture and serious thought. The first decisive turn, it is suggested, was taken by Galileo, in the attempt to subject all of nature to mathematics. This, by stages, led eventually to the prejudice that the only things that truly exist are those that can be weighed, measured or counted in some way. Against this background, the 'lived world' is displaced and value and meaning can no longer find purchase. Philosophy, says Husserl, becomes a struggle for its own existence (Husserl 1970: 13).

An enormous number of questions arise from this account, and the purpose of this collection is to pose and consider some of them. Is Husserl correct in his assessment of the scientific ambition since Galileo? Did earlier scientists or philosophers have the totalising ambitions for science, or a particular science, so tellingly set out by Quine in his response to Goodman's claim that science – and, in particular, physics – is only one version of the world, one 'way of worldmaking'? Quine remarks:

> Why, Goodman asks, this special deference to physical theory? This is a good question, and part of its merit is that it admits of a good answer. The answer is not that everything worth saying can be translated into the technical vocabulary of physics; not even that all good science can be translated into that vocabulary. The answer is rather this: nothing happens in the world, not the flutter of an eyelid, not the flicker of a thought, without some redistribution of microphysical states. It is usually hopeless and pointless to determine just what microphysical states lapsed and what ones supervened in the event, but some reshuffling at that level there had to be; physics can settle for no less. If the physicist suspected there was any event that did not consist in a redistribution of elementary states allowed for by physical theory, he would seek a way of supplementing his theory. Full coverage in this sense is the very business of physics and only of physics.
>
> (Quine 1981: 98)

But can such a scientific worldview co-exist with other accounts, or does it in principle or by tendency eliminate them as illusions?

Most of the papers in this collection were presented at a seminar series organised by the Philosophy Programme of the University of London School of Advanced Study in 1996–7, as part of its annual *History of the Problems of Philosophy* seminars. The idea of the series is to trace a problem of philosophy, through its variation in formulation, approach and attempted solution, from the ancient world to the present day. 'The Proper Ambition of Science' may not (yet) be the name of a classic problem of philosophy, but it is a question that arises in many guises and at many times. While this collection may well be one of the most comprehensive attempts to explore these issues, complete coverage unfortunately is not possible and it is not surprising that the majority of the papers here are concerned with modern and contemporary accounts. Even there the treatment has to be partial and, for example, little in this volume concerns the work of Hume and Kant. This we regret, but there are always limits to what can be accomplished in a single volume.

The collection begins with R. W. Sharples's paper 'Science, Philosophy and Human Life in the Ancient World'. Sharples centres on the question of whether any ancient thinker can be seen as proposing an eliminative understanding of science; that is to say, a vision of science which not only aims at universal understanding, but one which claims that non-scientific explanations are illusory and have no place in a fundamental account of the world. Although the philosophical concerns and doctrines of antiquity often seem very close to our own, they can also seem very distant. Among the ancients only certain Stoics imagined that any science could aim at complete coverage, and then only astrology. Interestingly, though, this doctrine was not based on any idea of the causal priority of the heavenly bodies, but on the causal inter-dependence of all things: by modern standards a view which deserves somewhat more respectful appraisal. However Sharples concludes that neither in astrology nor in medicine – his other leading example – did any ancient thinker rigorously propose an eliminative view even within the scope of its own sphere of application. Even ancient atomism appears not to be eliminative.

We move on the thirteenth century, and in order to bring out some of the major concerns of 'high medieval science' M. W. F. Stone concentrates on Albert the Great's writings on the hierarchy of the sciences. Stone explains the nature and development of Albert's subtly shifting views, situating his thought in the context of a revival of interest in all facets of Aristotelian philosophy, and the engagement of that tradition with Neoplatonism. One thing remains constant through the changes in Albert's view: theology – the contemplation of the divine – stands fast at the top of the hierarchy, while the natural sciences, despite Albert's immense regard and knowledge for his time, remain at the bottom. This does not entail that 'science' is essentially inferior to theology. Rather, the point for Albert is that all spheres of learning are essentially connected. Theology provides the individual sciences with their point and purpose.

G. A. J. Rogers continues the theme of the relation between the established teaching of the church and the growth of scientific knowledge as they increasingly came into tension in the seventeenth century. However just as central to Rogers's account is the conflict between the new atomistic theories of matter and traditional epistemology once it is appreciated that on the atomistic theory we do not see things as they really are in themselves. This is the beginning, then, of the dislocation between the scientific world view and our intuitive accounts of the 'lived world' that so troubled Husserl. Rogers, however, reminds us of a point that complicates the picture. Few, if any major seventeenth century scientific philosophers thought that certainty could be achieved in the natural sciences. Thus any particular scientific world view is only, on this view, a theory.

J. R. Milton takes up another aspect of the thought of this period: the assault by both scientists and philosophers on the notion of a hierarchy of degrees of perfection. Modern science is generally seen by both its advocates and critics as value-free: older concepts of perfection and nobility that appear to straddle the fact/value divide are firmly rejected. Milton describes the abandonment of hierarchical concepts in both physics and metaphysics during the course of the seventeenth century, and discusses the possible connections between these changes, arguing that neither change should be seen simply as a consequence of the other.

With Aaron Ridley's discussion of Nietzsche we move to more contemporary concerns. Husserl was not alone in his concern about increasing scientism (indeed many of Husserl's claims echo those of Nietzche). As Ridley says, 'Nietzsche is acutely aware of the scientism of his contemporaries, and had he had us for his contemporaries his awareness would have been acuter still.' Yet as Ridley demonstrates, science for Nietzsche in its proper place – acknowledged as one perspective among others rather than as a privileged perspective-free vantage point – is in 'the service of life'. Scientism, by contrast, stunts life by forcing it into a single stultifying pattern.

Christopher Hookway considers how these themes are worked out in the writings of the pragmatists. The simple picture is that Peirce adopts some form of singular scientific method in philosophy, while James is a pluralist, privileging no version of the world above any other. Thus Hookway quotes Peirce in terms that would apparently make Husserl wince; as announcing his intention to bring 'modern mathematical exactitude to philosophy' and to 'rescue the good ship Philosophy for the service of Science from the lawless rovers of the sea of literature'. Yet even Husserl may agree with Peirce's underlying meaning; that philosophy should be conducted with an intense desire to discover the truth.

We must ask, though, whether science is to be characterised by a subject matter (Dewey) or a method (Peirce), which raises the further question of whether the same subject matter can be approached in both a scientific and non-scientific manner. But as Hookway shows, while there are important differences of emphasis among the pragmatists, attempts to line them up on one side or other of a 'scientistic' and 'pluralist' divide leaves out much of interest in all views. Furthermore little Peirce says should provide any comfort for the modern day 'naturalist' who wishes to base philosophy on the results of the special sciences.

We began in this introduction by setting out some of Husserl's concerns about the increasing dominance of the scientific methodology upon European thought. Thus it is something of an initial surprise to see that, according to Dermot Moran, Husserl not only was not opposed to science but that he 'saw science as the only hope for the salvation of humanity', and, indeed, thought philosophy should live up to the ideal of itself as a rigorous science. This scientific model of philosophy was proposed as a protective against irrationalism or relativism, yet it was equally important for Husserl to avoid naturalism and scientism. Moran points out that for Husserl 'true objectivity is found not by excluding subjectivity but precisely by taking it into account'.

Thomas Uebel takes us into the world of logical empiricism and to a highly specific case study in the philosophy of social science. Both Popper and Hayek argued against what they saw as a massive misconception of the possible application of scientific ideas: the political reorganisation of society upon 'scientific' lines. Uebel suggests that, despite the differences between them, both Popper and Hayek also aimed their critiques at an unspecified enemy: the philosophy which they thought stood behind Otto Neurath's proposals for rational economic planning. As Uebel maintains, on a reading of the texts it is far from clear that Neurath deserved such criticism. Nevertheless, in the process of examining Popper and Hayek's charges we come to understand that many logically distinct positions can be called 'scientism' and if, indeed, they are all mistaken, then they are not all mistaken in the same way.

The collection ends with two papers arguing for different accounts of the proper ambition of science, or, again, at least of physics. David Papineau considers the doctrine of contemporary physicalism, which he defines as the claim that everything is physically constituted: an ontological doctrine. Like Quine he sharply distinguishes this from the methodological doctrine that everything should be studied by the methods of the physical sciences. Contemporary physicalism, Papineau argues, is motivated by a simple argument, based on a premise which, he claims, became available – or at least commonplace – only in the twentieth century: the premise of the completeness of physics. This is the thesis that all physical effects are due to physical causes rather than, say, that some are due to 'vital motions'. (This is to be understood as to leave open questions about any sphere which may not have physical effects, such as the moral or mathematical.) Papineau traces the chequered history of the theory of the completeness of physics through post-Galilean mechanics. His conclusion is that, understood correctly, it has now been established, by any reasonable standard, by more than a hundred years of detailed empirical research.

Nancy Cartwright reads the empirical record another way. Her understanding of the thesis of the completeness of physics is phrased in slightly different terms – that the laws and theories of physics can in principle subsume everything – and she is highly sceptical. Not only does Cartwright dispute that physics can account for everything outside its immediate domain, she claims that 'physics cannot account for everything that is in its domain'. The mistake can, at least in part, be attributed to our tendency to overlook the fact that physics enjoys its extraordinary predictive success only in those areas where *ceteris paribus* laws frame the domain under consideration. Outside such 'nomological machines' causes which are not part of the system will exert disruptive effects. As an alternative to the 'pyramid' of reduction with physics at its tip, Cartwright offers the image of a patchwork of laws governing a 'dappled world, a world rich in different things, with different natures, behaving in different ways'. But as Cartwright warns, this is simply a different image, and neither the image of a pyramid or a dappled world should be allowed to dominate our thinking.

What then may we conclude from these explorations? Perhaps only that the views of every major thinker are more subtle, more nuanced, and, perhaps, more reasonable, than we come to think if we attend only to the stark oppositions in which they are so often presented. But on further reflection, interesting conclusions can be drawn. From ancient times to the present day the question of how the body of knowledge demarcated 'science' relates to the nature and scope of philosophy has been of pressing concern. From ten particular case studies we see that how this concern presses differs from age to age and from thinker to thinker, but press it does. So while there is no unity on what, exactly, the problem is – still less on how it should be resolved – we see continuous engagement spurred on by the belief that there is something problematic in this area. But we should not expect cosy consensus on the nature of the problem and its solution to break out any time soon. For evidence one need only consult the final two papers in this collection.

Bibliography

Husserl, E. (1970) *The Crisis of European Sciences and Transcendental Philosophy*, trans. D. Carr, Evanston: Northwestern University Press.

Quine, W. V. (1981) 'Goodman's *Ways of Worldmaking*' in his *Theories and Things*, Cambridge Mass.: Harvard University Press.

2 Science, philosophy and human life in the Ancient World

R. W. Sharples

Science and understanding

How far did anyone in Greco-Roman antiquity anticipate contemporary claims for the *universal* authority of science? What, indeed, is involved in even asking this question where the ancient Greco-Roman world is concerned? To begin an essay by stressing the importance of defining the terms in question can produce a rather stilted and mechanical effect; but sometimes it is necessary.

'Science' is not only an anachronistic category where the ancient world is concerned, but an insular, or at least Anglophone one, even today. Other modern languages do not limit the scope of 'science' in the same way that English does. The study of classical antiquity itself is in German *Altertumswissenschaft*, and it is sponsored in France by the Centre Nationale du Recherche *Scientifique*. True, the labelling of classical studies as *Altertumswissenchaft* reflected a deliberate attempt in nineteenth-century Germany to present the subject as 'scientific'; an attempt in which English classical studies eventually followed Germany, but reluctantly and with many wistful backward looks in the direction of art, in the English cultural sense of the term, as opposed to science. However, the native German speaker who pointed this out to me then unintentionally confirmed my view that we *were* after all dealing with conceptual differences as well as with persuasive definitions, by going on to ask, in all sincerity, whether we did not in this country naturally describe theology and religious studies as 'religious science'.

Myles Burnyeat has pointed out that the ancient Greek word we normally translate as 'science' or 'scientific knowledge', *epistêmê*, is better rendered as 'understanding', a translation that no one has ever hesitated to use for the cognate verb *epistasthai*. One can *understand* why rain falls, or why the old are less cheerful than the young; or at least, as we shall see, Aristotle thought one could. It is at least questionable whether one can in the same sense *understand* why Beethoven chose the themes he did for the

Seventh Symphony, or why a student may come to a regular lecture on some occasions but not on others. At least, Richard Sorabji has used the latter as an example of the sort of thing that may have no explanation, (Sorabji 1980: 30–1) and if it has no explanation we presumably cannot understand it either.

This might suggest a rephrasing of our initial problem in terms of the question: 'how far did people in antiquity suppose that the world in general, and things that affect human beings in particular, could be *understood*?'[1] And one could at this point cite well-known passages like one in Philoponus that stresses the limits of human, as opposed to divine, understanding.[2] One could also allude to E. R. Dodds's famous attempt, in *The Greeks and the Irrational* (Dodds 1951: especially chapter 8), to show not only that the ancients were much less 'rational' than some had tended to think, but also that rationality had gained ground in the fourth century BC, in particular, only to be driven back by renewed tides of superstition in the period of the Roman Empire. However, if one chose to adopt this approach one would need to bear in mind that the boundaries between the rational and the irrational, between reason and superstition, have not always been drawn where we would now draw them, or at least where some of us would now draw them. For the Stoics in the third century BC the prime examples of studies which showed the inclusion of all phenomena in a nexus of cause and effect, and which demonstrated how a science could be based on observed correlations even if the reasons for these were not understood, were astrology and divination from the entrails of sacrificial victims.[3]

The issue of eliminativism

It seems better to adopt a different approach, and to start, as Aristotle puts it, from what is most knowable to us rather than from what is most knowable in itself. In other words, what is the contemporary agenda that is prompting the question this volume of essays is addressing? For I am firmly convinced that we cannot, in studying classical antiquity, escape from our own preoccupations; the important thing is to be aware of the way in which they are influencing us, and not to import anachronisms without being aware that we are doing so. And I take it that the underlying question has to do with science whose claims are not just reductivist but eliminativist; that is to say, science that claims that explanations in terms other than its own are ones that we would be better off without. To be sure, this question cannot in practice be separated from that of just what sorts of explanation are scientific and what are not. Is psychoanalysis to count as scientific, for example?

We can certainly find in antiquity the view that everything can be explained in terms of a single unified system of 'understanding'. Plato notoriously regards geometry as limited because it does not treat its fundamental assumptions as themselves deriving from the nature of goodness (Plato, *Republic* 6 510cd). But while a notion of 'understanding' that is so comprehensive may be seen as misguided, and was so already by Aristotle, who insisted against it that what different fields of enquiry shared was their methodology, not their starting-points or their subject-matter (*Posterior Analytics* 1.7, 1.9–10), it can hardly be seen as eliminativist.[4]

What I rather want to concentrate on is the question of what I would like to label 'aggressive eliminativism', the claim that a science that some would see as essentially *limited* in scope in fact supersedes other modes of understanding. I do not therefore want to look at the fifth-century debates over rational medicine and magical healing;[5] for that is, in some sense, a matter of medicine and magic disputing the same territory, not of a recognised science seeking to expand beyond its traditional territory. Debates in the fifth century BC about the divinity or otherwise of the heavenly bodies might provide a closer parallel; but even though these were seen at the time as challenging traditional religious belief,[6] and even though there were at the time people such as Protagoras who regarded the gods as irrelevant (fr. 4), claims that the heavenly bodies were not divinities need not in themselves imply the elimination of all religion of whatever sort. My concern is rather with scientific claims that do not just restrict the scope of other modes of understanding but actually claim to eliminate them altogether. A closer parallel might be Chrysippus' insistence, following Socrates, that knowledge is sufficient for virtue, and that undesirable emotional states reflect, or rather simply *are*, faulty judgments about what is good or bad in our situation. For Chrysippus' view was seen by Galen as a rejection of Plato's recognition that human beings have an emotional as well as an intellectual nature.[7] Or, putting it another way, for Galen, we might say, Chrysippus is extending the bounds of rational argument too far; he is handing over to logic those regions that should be the preserve of what we would now call psychology. True, both for the ancients and for us (perhaps) both logic and psychology may be seen as in some sense scientific, so that we are dealing with a demarcation dispute between sciences rather than with a question about the scope of science altogether; and the attempt at expansion did not come from the direction we might expect it to come from today.

In the Aristotelian tradition, explanations that we would think of as physical, and those which we might not, coexist without apparent tension. Anger is, in material terms, the boiling or seething of blood around the heart; but, just as for Aristotle we cannot understand the structure of a living creature if we do not consider it in terms of purpose, just so we will not understand what anger is if we consider only its material aspect.[8]

Plutarch's seeming inconsistency over whether the change of colour of the octopus was for Theophrastus a reflex reaction to danger, or something done for the sake of self-protection, can be explained as a failure on his part to recognise that, for an Aristotelian, at some levels of animal behaviour the contrast between what is automatic and what is purposeful is simply inappropriate; in a teleological Aristotelian universe animals simply are structured so as to do the appropriate thing (cf. Sharples 1995: 41, 93–6). At the human level, however, it is difficult to avoid questions of priority as between physiological and psychological explanations; and sometimes the connection between the two seems unclear, as when Aristotle in *Rhetoric* 2.13 explains the generally negative attitudes of the elderly both by their experience of life *and* by their colder constitution.[9] The explanations may not indeed be incompatible, but one would like to hear more about the relation between them.

The idea of dual explanations is one that has figured in discussions of the claims of science from antiquity to the present day; a famous ancient example is Plutarch's account of the portent of the one-horned ram.

> It is said that once a ram with only one horn on its head was brought to Pericles from the country, and Lampon the diviner, when he saw that the horn was strong and solid and grew from the middle of the forehead, said that there were two power groups in the city, that of Thucydides and that of Pericles, but the power would come into the hands of one person, the one to whom the sign had been given. But Anaxagoras cut the skull open and showed that the brain had not grown to fill its place, but was pointed like an egg, sliding together from the whole container into the place where the root of the horn started. And at the time all the bystanders were impressed by Anaxagoras, but shortly afterwards by Lampon, when Thucydides' power was broken and all the affairs of the people were uniformly under the power of Pericles.
>
> *But I think there was nothing to prevent both the naturalist* (phusikos) *and the seer being right, the one grasping the cause well, the other the purpose*. For it was the task of the one to consider from what the thing resulted and how it came about naturally, but for the other to predict with what end in view it came about and what it indicated. Those who say that finding the cause does away with the significance do not realise that they are rejecting artificial symbols along with divine ones, like the sound of gongs and the light of beacons and the shadows of sundials. Each of these has been made to be the sign of something by some cause and contrivance. But perhaps these things belong to another treatise.
>
> (Plutarch, *Life of Pericles* 6.2)

Plutarch, a Delphic priest as well as a Platonist, had a particular interest in maintaining the validity of religious as well as of naturalistic explanations; and the passage is evidence for the existence of eliminativist claims on behalf of the latter. For if such claims had not existed, Plutarch would not have felt the need to counter them.

Eliminativism is more clearly at issue in the Atomist tradition. Democritus in the fifth century BC, attempting to formulate a physical theory that would take account of the arguments of Parmenides, claimed that sense-experience is not simply illusory, as Parmenides supposed, but can be accounted for in terms of an underlying reality. However, he followed Parmenides in holding that reality was only accessible to reason, while the evidence of our senses puts us in direct contact only with appearances. By convention there is colour, or this table; in reality there are only atoms and the void (Democritus, fr. 9). However, it does not seem that Democritus saw this epistemological and metaphysical claim as having practical implications for our understanding of the world as we experience it. Our evidence for Democritus' ethical and political thought is uncertain and secondhand, but the consensus seems to be that, while it was naturalistic in the sense of not *conflicting* with his atomic physics, it was not aggressively dismissive of conventional values and customs (Guthrie 1965: 495–6).

David Sedley has shown that the atomism of Epicurus in the late fourth century BC, by contrast, was strongly *anti*-reductionist; the table and its colour are as real on their level as the underlying atoms are on theirs.[10] In interpreting Epicurus thus, Sedley was in effect reviving the claim of Cyril Bailey earlier in the present century that Epicurus' association between the random swerve of atoms and rational human choice was to be explained by the claim that what was merely a random swerve on the level of individual atoms took on a different character when it occurred in a complex compound such as a human being. What Sedley has done is to link this view of Bailey's to a systematic account of the difference in emphasis between Democritus and Epicurus and of the historical context in each case.[11]

Epicurus, notoriously, explained the phenomena of life and consciousness by arguing that the make-up of the human soul and mind – the soul for him being a physical part of us with functions not altogether unlike those which we would attribute to the nervous system – included atoms of an otherwise unknown type. This has been described as stretching materialism to its utmost limit (Bailey 1947, vol. 2: 1027). What has perhaps less often been emphasised is that it is in conflict with Epicurus' own principles; for the point of the atomic system, and one that Lucretius repeatedly emphasises, is that compounds of properties can have properties that individual atoms do not. Indeed Lucretius in his second book is at pains to emphasise that animate beings can be made up of inanimate particles (Lucretius, 2.866 ff.), and that we need not suppose that, because we can laugh, the atoms of which we are made can do so too. It is therefore distinctly odd when we go on, in the following book, to read that life and consciousness require the presence of atoms that are indeed inanimate in themselves, but different in kind from any others (Lucretius, 3.228 ff. Cf. on this Sharples 1991–3: 189–90).

Science, philosophy, literature and culture

Both Epicureanism and Stoicism are 'scientific' philosophies, in the sense that they claim that bodies are the only realities and that everything can be explained in terms of those bodies. (For Epicurus a compound body is a body as much as an individual atom is.) Epicurus, like Democritus before him, related the properties of compounds to the shapes and sizes of their constituent atoms; and the Stoic Chrysippus asserted that for every difference in outcome there must be a reason:

> Speaking against these people on the grounds that they violate nature by [introducing] what is without cause, Chrysippus in many places adduces the knucklebone and the balance, and many of the things that cannot [he says] fall or incline in different ways at different times without some cause and difference either concerning them or concerning what comes from outside. *For what is uncaused and spontaneous is altogether non-existent*; in the [case of the] adventitious impulses that are invented and spoken of by certain people undetected causes intrude, and we are not aware that they lead the impulse in one direction [rather than the other].[12]

However, Democritus' and Epicurus' concern was with stating the principle of the connection between properties and atomic shapes, rather than with investigating the details and refining the theory; and Chrysippus' claim that nothing happens without a cause was a theological claim, concerned with the unity of the divinely ordered universe, rather than the prelude to a research proposal. It is true that Chrysippus claimed that everything can be explained in physical terms, by the tension of *pneuma* or 'spirit', a mixture of expanding fire and contracting air present in every thing; this was certainly the promotion of a physical principle to the role of universal explanation, but just because of its universality and the lack of detailed development of the theory it does not so much threaten other modes of explanation as subsume them, rather as the Stoic claim that God governs everything in the world for the best tends in practice, if not explicitly, to become the claim that whatever happens is good because it happens. There was a traffic in ideas and terminology between Stoic philosophy and sciences such as medicine; but that is not the same as Stoic philosophers themselves engaging in scientific enquiry, and the one who notably did, Posidonius in the first century BC, is described for this very reason by Strabo as 'Aristotelianizing' (Strabo, 2.3.8 = Posidonius T85 Edelstein-Kidd).

It is indeed characteristic of ancient culture to use ideas that we might call scientific for other, literary or rhetorical, ends.[13] An extreme case, for which I am grateful to my colleague Maria Broggiato, is in an interpretation which even the allegorist Heraclitus Homericus describes as 'monstrous' (the Greek word is *terateia*: *Homeric Questions* 27. 2–4). Crates of Mallos explained Zeus' hurling Hephaestus from heaven in the *Iliad* by his desire to use the time of Hephaestus' fall to measure the size of the universe: *Zeus anametrêsin tou pantos espoudakôs*, 'Zeus having become eager for a measurement of the universe' (Crates fr. 22a Mette 1936). Homer, after all, tells us (*Iliad*: 1.591 ff.) that Hephaestus took a day to fall to earth; clearly then Zeus's motive for his injurious act, which left Hephaestus, as a god, lame for eternity, must have been the disinterested pursuit of scientific enquiry. Crates further made the point that the reference to a day shows that the movement of one fiery body, the sun, was used to time that of another, Hephaestus. It seems more than doubtful that we should draw any inferences from this about what Crates thought Zeus's actual results were; to do so would after all involve assumptions about the rate of acceleration of a falling Hephaestus.[14]

This is an extreme case; but it has its basis in the fondness of the Hellenistic period for the strange and exotic and for the challenge of incorporating it into traditional forms, the sort of attitude that rapidly turned the enquiries of the Lyceum into the pseudo-Aristotle *Mirabilia*, and that led to Nicander's elegant versification of remedies for serpent-bites, or later on to Oppian's poem on fish.

Such an attitude indeed presupposes some familiarity with, or receptivity to, the science of the day among the cultural elite who would provide the audience for literature. Where one science, medicine, is concerned Alberto Jori has shown how Aristotle in the *Politics* (1282a 3–7) recognises a class of people who are not themselves practising physicians, but know enough to be able to assess the performance of the professionals, thus providing a degree of consumer control, or, as he also notes, of aristocratic counterbalance to a developing professionalism (Jori 1995. See also Jaeger 1945: 10–16). Jori links this with what is said about the person of general culture at the beginning of Aristotle's *On the Parts of Animals*:

> Concerning every study (*theôria*) and investigation (*methodos*), alike with what is more humble and with what is more honourable, there seem to be two types of disposition. It is appropriate to call one of these understanding (*epistêmê*) of the matter, the other a sort of being-educated (*paideia*). For it belongs to the educated person's character to be able to judge accurately what the person who is speaking explains well, and what not. For this is something like what we suppose the generally educated person to be, and to have been educated is to be able to do what has been said: except that we think that this person is a single judge of everything, so to say, while the other (the expert) (can judge) about a single definite nature.
>
> (Aristotle, *Parts of Animals* 1.1 639a1 ff., with acknowledgements to Peck's Loeb translation)

In the Roman empire too some knowledge of medicine was a part of general culture among the elite. True, medicine may be a special case, or at least at one end of a spectrum; it is, after all, an intensely practical matter where people might well feel – and not just in the conditions of the ancient world – that not everything should be left to the experts.

Science, morality and literature: the case of Seneca

The Roman writer Seneca was a Stoic. He thus believed that the world was the expression of a divine plan; human beings form part of that world, and 'living in conformity to nature' was the Stoic definition of the goal of life. One might therefore expect that in his *Natural Questions* he would not only see the study of nature as the study of divine workmanship, as suggested by Gross 1989, but would also derive from it conclusions about how human life should be lived, in the tradition of Stoics who, from the observation of new-born creatures, argued against Epicurus that self-preservation, rather than pleasure, was the primary natural instinct (Diogenes Laertius 7.85–6 = SVF 3.178 = Long and Sedley 1987, 57A. Cf. Brunschwig 1986), or who interpreted the traditional cardinal virtues as originating in natural human instincts (Panaetius, reported by Cicero, *De Officiis* 1.15).

Seneca does indeed draw moral lessons from the observation of nature, but not in quite the way one might have expected. True, in the *Natural Questions*, and not only there, he appeals to nature to argue against what he regards as unnatural human behaviour.[15] But the connections he draws between nature and human behaviour are not always the most obvious or straightforward. For Seneca was a rhetorician; his writings are characterised by the cleverness for cleverness' sake characteristic of the so-called Silver Age, and part of the point of the *Natural Questions* is the ingenious nature of the links made between one topic and another. It is not for nothing that Ovid's *Metamorphoses* is so frequently quoted in this work of Seneca; for it too is a work characterised by the uniting of apparently disparate elements into a single whole, and by the cleverness of its transitions. Not that this stops Seneca from criticising Ovid for trivialising the Great Flood by having wolves and sheep swimming together; this, Seneca says, is 'childish silliness', since they would all have been drowned anyway. As Seneca sternly – on the face of it, at least – remonstrates, 'It is not a sufficiently serious attitude to make fun of the whole world now swallowed up' (*Natural Questions* 3.27.13 ff.).

The discussion of the rainbow, which involves consideration of reflection, leads by a natural progression – natural for Seneca, anyway – to the denunciation of Hostius Quadra, who used magnifying mirrors to get a better view of himself while engaged in unnatural vices (1.16). This, Seneca argues, was a misuse; in a world governed by purpose, the function of reflection – natural reflection in water, as well as artificial reflection in man-made mirrors – is to enable us to contemplate ourselves and thus gain self-knowledge and wisdom, so that the ugly can know that virtue is more important than being handsome, and so that the elderly can have proper respect for their own grey hairs (1.17.4–5). Both lightning (2.59) and earthquake (6.32) prompt thoughts of the folly of fearing a particular form of death when we are all under a sentence of eventual death in any case.[16] The discussion of snow leads on to an attack (4b.13.3 ff.) on those who buy snow and ice to chill their drinks, only needing to do so because of their unhealthy and indigestible diet; in doing so they are buying water, which (Seneca says) should be free to everyone, and moreover water mixed, in the case of snow, with a large quantity of air.

True, the connections Seneca makes between natural phenomena and moral lessons, even if ingenious, do still rest on the Stoic assumption that the world is a single unified system of which human behaviour is a part like any other, and we are not yet at the point where a late-medieval compendium for preachers, the *Lumen Animae* or *Light of the Soul*, could arrange its examples from natural science – a few attributed, on who knows what grounds, to Theophrastus – by the morals they served to indicate, rather than by a taxonomy derived from the natural world.[17] The *Lumen* appears to claim, for example, that both the flying up of sparks when two burning sticks are struck together, *and* the alleged uniting of two pieces of rotting wood underwater into a single petrified mass, illustrate the reconciliation of the Virgin Mary and the sinner which is only possible through the moisture of devout tears (*Lumen Animae B 7*, *De Beata Virgine*). I do not understand how that is meant to apply to the fire case.

Nor do I want to suggest that Seneca is lacking in genuine scientific spirit.[18] The moralising passages are a relatively small part of the whole; and Seneca can write expressively, indeed movingly, of the limitations of present knowledge, and of the greater understanding of nature that will be possessed by future ages when Seneca and his contemporaries have been forgotten (7.30.5). This too is rhetoric, but then *everything* in Seneca is rhetoric, and if we discount it on those grounds we will not understand him. And he ends by giving an explanation of the limited understanding of his own time; there is no interest in philosophy, but plenty in the pantomime and in gladiators (7.32). 'Who respects a philosopher, or any liberal study, except when the games are called off for a time or there is some rainy day which he is willing to waste?' (7.32.1). Some things – and some complaints – do not change.

Astrology, medicine and morality

However, appeals to science in literary and rhetorical contexts, showing that scientific interest is a part of wider culture, are one thing; claims by science itself to universal competence, or at least to wider competence than might initially be allowed, are another. And I wish to conclude this paper by considering two examples, one from astrology – in ancient terms not distinct from astronomy – and the other from medicine.

It is sometimes claimed, by modern defenders of astrology, that whereas ancient astrologers claimed to foretell every occurrence, modern astrology only indicates general tendencies, thus preserving free will. In fact, the situation is less clear-cut. We do indeed in antiquity find Favorinus criticising astrology for extrapolating from the admitted effect of the heavens on such phenomena as the tides to claim that they affect the most trivial details (Gellius, *Attic Nights* 14.1.3), and criticising astrology for submitting not only external events but our own choices to the control of the stars (*Attic Nights* 14.1.23). The greatest astronomer of antiquity, Ptolemy, however claims that the nature of the sublunary world prevents everything being determined precisely by the movements of the stars, and that room is thus left for human agency (Ptolemy, *Tetrabiblos* 1.3.7. Fazzo 1988: 638). Ptolemy's position here has been seen as the source of Alexander of Aphrodisias' theory of fate as natural tendencies that can be resisted, though the extent to which there is even tacit reference to astrological ideas in Alexander's discussion is controversial (Sharples 1983: 18–19, 128; Fazzo 1988). Alexander certainly alludes to astrology as a popular belief in support of his own identification of fate and nature, since the movements of the heavenly bodies are in his own view the first causes of natural coming-to-be (Alexander, *On Fate* 6: 169.23 ff.); but that is different from endorsing astrology himself. We also find Alexander arguing – perhaps indeed only as a debating point – that the indeterminacy in things explains why divination does not always foretell the future accurately; while the Stoics, using divination as evidence for the rigid interconnection of all events, explained failures in divination by the shortcomings of the practitioners.[19] Alexander supposed that medicine is essentially stochastic or imprecise – as we might say, a hit-or-miss affair – whereas Galen had supposed that its shortcomings reflected the gap between general theoretical rules and their practical application to each specific case.[20]

By contrast with Ptolemy, Manilius, who wrote a didactic poem on astrology in the first part of the first century AD, presents the claim that *everything* is determined by the stars as a ground for freedom from anxiety.

> Set your minds free, mortals, and lighten your cares; empty your life of so many superfluous complainings. The fates rule the world, all things are fixed by a firm law and the long intervals of time are marked out by fixed happenings. Our death is in our birth, our end depends on our beginning. It is from this that power and wealth flow and, more often, poverty; that skills and character are given to those who are created and faults and praises, losses and gains. No-one can be without what has been given to him or possess what has been denied, (no-one can) grasp fortune by his prayers when she is unwilling or flee from her when she threatens; each must endure his own lot.
>
> (Manilius, *Astronomicon* 4.12)

Moreover, he defends the legitimacy of punishment in a way that explicitly eliminates considerations of *moral* responsibility altogether.[21]

> Finally, if the order of fate does not exist, why is it handed down, and all that will come to pass at fixed times prophesied? – Nor yet does this argument defend crime or deprive virtue of its rewards. *For no-one hates poisonous plants less, because they come not from choice but from a fixed seed; nor is less gratitude shown for pleasant food because it is nature that produces the crops and not any will.* So let the glory of men's good deeds be so much the greater because they owe the praise to heaven, and again let us hate the guilty the more, as they were created for guilt and punishment. Nor does it matter what the origin of the crime is; it must be admitted to be a crime. This also is fated, to pay the penalty for what is fated (or: this too is fated, thus to expound fate?).
>
> (Manilius, *Astronomicon*: 4.106–18)

This *may* indeed be rhetorical bravado, the defending of an extreme case.

The question arises how astrology, if it applies to everything, relates to other types of explanation. At least in the orthodox Stoic view, astrological and other divinatory signs are not *causes* of what they portend; rather, both are linked together in the universal causal nexus (Cicero, *On Divination*: 1.117-18 = SVF 2.1210 = Long and Sedley 1987: 42E). What this would logically seem to imply is not that other types of causal explanation are invalid; rather, they are *required* to explain why the predictions of astrology are successful, and that is why the alleged successfulness of divination can be used as an argument for universal causal determinism. The implication is rather that for practical purposes of predicting the future astrology is, in principle, all we need.[22]

To regard responsibility as irrelevant in the way Manilius does is unusual in the ancient world.[23] It is *not*, in particular, the attitude of orthodox Stoicism; Chrysippus was concerned to argue that responsibility is compatible with determinism,[24] while Manilius holds that the wicked should be punished whether responsible or not. (This is not indeed to say that Manilius was *himself* conscious of the difference between his position and the orthodox Stoic one.) Manilius' position is however paralleled, as Theiler and Donini have pointed out, in a passage from the doctor Galen. (Theiler 1946: 56; Donini 1974: 146f) This is in the context of a rather different theory, the claim that the functioning of the soul is influenced by that of the body. It comes in the course of Galen's treatise entitled *That the powers of the soul follow the temperament of the body*.

> This argument does not do away with the good results of philosophy, but guides and instructs [us towards them], even though it is to an extent unknown to some philosophers. For those who think that all human beings are capable of virtue and those who say that no one chooses justice in itself have both only half understood human nature. For neither are all naturally hostile to justice nor all friendly (to it); and each group come to be like this on account of their bodily temperament. 'How then,' they say, 'could someone be justly praised or blamed or hated or loved, when he has become bad or good not because of himself but because of his temperament, which he clearly derived from other causes?' 'Because,' we shall say, 'this is natural for all of us, to welcome and admit and love what is good, and to avoid what is evil, not considering whether it is innate or not. *We destroy scorpions and poisonous spiders and vipers, which were made like this by nature and not by themselves. Reasonably therefore we hate wicked human beings, too, not taking any account in addition of the cause that made them like this*, and conversely we welcome and love the good, whether they came to be like this from nature or from education and

instruction or from choice and training; and we kill the incurably wicked for three reasons, reasonably so: in order that they may not wrong us if they live, in order that they may make those like them afraid of being punished for their crimes, and thirdly because it is better for them themselves to die when their souls are so corrupted.

(73.3–74.19 Müller (Leipzig, Teubner, 1891; 814–16 Kühn)

Galen presents his argument not only, or even primarily, as justifying punishment and reward, but as justifying praise and blame. Here the medical analogy might seem to break down: we may take steps to cure someone who is ill, or to protect ourselves from someone who is infectious, but normally without *blaming* them. The answer must be that Galen treats not only punishment and reward, but praise and blame too, purely as instrumental in altering behaviour. (The fact that we may blame *some* of those who are ill, those who are responsible for their own illness – cf., of physical shortcomings more generally, Aristotle, *Nicomachean Ethics* 3.5 1114a21–9 – is irrelevant to Galen's position, for to appeal to it as a parallel would involve a distinction, between cases where we are responsible and those where we are not, which his position does not in fact require and is more consistent without.)

Geoffrey Lloyd appears to interpret this passage as claiming that Galen defends praise and blame on the grounds that everyone can embrace good and reject evil, and thus avoids a completely determinist position (Lloyd 1988: 37). But it seems clear, from the immediately following references to destroying venomous animals and hating wicked human beings, that the reference of 'to love what is good and to avoid what is evil' is not to the agents' *own* actions, but to our natural treatment of *others*, the righteous on the one hand and wrongdoers on the other, regardless of whether or not we regard them as responsible for their actions.[25] It is true that on any account there is an inconsistency between 'this is natural for all of us, to welcome and admit and love what is good, and to avoid what is evil' and the statement just before that 'neither are all naturally hostile to justice nor all friendly (to it)', and more generally between what Galen says here and the importance he attaches to innate tendencies to evil as well as to good elsewhere in the treatise. (Cf. Donini 1974: 146; Lloyd, loc. cit.)

The point that praise and blame do not involve responsibility is logically connected with the effect on our behaviour of our original, innate and inherited, constitution. But Galen also wants to insist that, if our judgement and our behaviour are affected by our bodily temperament, and that in turn is affected by our diet, then in order to be virtuous we should follow the right diet. Indeed shortly before the passage quoted Galen argues that by *trophê* in *Timaeus* 87b Plato meant nourishment in the sense of food, rather than 'education':

> For the moment though I will write out the passage in the *Timaeus*, in which Plato, having first said 'in this way all of us who are bad become bad on account of two most involuntary factors. And for these the planters should always be blamed rather than the crop, and the nurturers than what is nurtured', goes on to say 'we must try, in whatever way anyone can, both through nurture and through our practices and studies to flee from wickedness and acquire its opposite.' For just as 'practices and studies' remove wickedness and produce virtue, just so does 'nurture'. *Nurture is sometimes spoken of by them*[26] *with reference not only to that involving food but also to the whole upbringing of children; but it is not possible to say that Plato now speaks of 'nurture' in the second sense.* For it was exhorting not children but adults that he said 'we must try, in whatever way anyone can, both through nurture and through our practices and studies to flee from wickedness and acquire its opposite.' So by 'practices' he means those in gymnastics and music, and by 'studies' those in geometry and arithmetic. *It is not possible to understand 'nurture' as anything other than that from (solid) food and porridge and drinks* – which include wine, concerning which Plato expounded a great deal in the second book of the *Laws*.[27]
>
> (71.17–72.18 Müller = 812–14 Kühn)

Earlier, stressing the connection between diet and behaviour in Plato, *Laws* 5 747d, he has challenged those who deny it to come and hear from him what things they should eat:

> For they will be benefited greatly as far as ethical philosophy is concerned, and in addition to this they will progress to excellence in the powers of the reasoning (part of the soul), becoming more intelligent and more able to remember.[28]
>
> (66.7–67.16 Müller = 807–8 Kühn)

And in the sequel to the discussion of the *Timaeus* quoted above Galen observes that:

> Whoever wants, apart from this, to learn something about the whole power of nutrition can read my three books about this and, in addition, as a fourth the one on good and bad humours, which is most necessary for the present topic. For a bad state of the humours greatly harms the activities of the soul, while a good one preserves them unharmed.
>
> (72.18–73.2 Müller = 814 Kühn)

On this Donini comments that (in Galen's view) 'science and medicine in a way have more power than traditional education or philosophical education, which we have already seen to be ineffective against certain (moral) perversions or natural shortcomings.' (Donini 1974: 144)

The claim that what we eat affects what we are is not unparalleled in antiquity. As Richard Sorabji has shown, the Neoplatonist Porphyry advocated vegetarianism for the philosopher not on the grounds of animal rights or of any moral argument, but rather because of the alleged effects on the intellect of the eating of meat.[29] As Shakespeare made Sir Andrew Aguecheek say in *Twelfth Night* (Act 1 Scene 3), 'I am a great eater of beef, and, I believe, that does harm to my wit'.

Plato himself had made Socrates in the *Republic* argue the need for a healthy body as well as a healthy mind. He insists (403d) that the soul makes the body good rather than the reverse; but he immediately goes on to describe how food and drink can have deleterious effects on intelligence. Galen in *That The Character. . .* attacks Platonists who hold that the body can hinder the mind but not benefit it;[30] he himself wants to insist on the positive as well as the negative effects of bodily temperament, and thereby is claiming a larger role for medicine in relation to human behaviour.[31] Whether Galen in our treatise commits himself to the view that medicine is the *only* science relevant to human behaviour, and excludes all others, is uncertain; so too is the question whether, if he does so, he does so intentionally. Ballester observes that the implication of Galen's discussion is to 'place the doctor and his activity at the top of all professional activities' (Ballester 1988: 129).

The nature of the soul is a topic on which Galen, both in this treatise and elsewhere, is explicitly agnostic; but his view in this treatise can be categorised as either eliminativist or (more probably) epiphenomenalist.[32] He objects to Andronicus' definition (later adopted by Alexander of Aphrodisias) of the soul as the *power that follows upon* the mixture of the bodily elements, and insists that it just *is* the mixture.[33] But Caston argues that this is not in fact Galen's true position, and that the title of the treatise represents Galen's position more accurately.[34] Eliminativism would imply that all human activity which is commonly regarded as psychological would in fact better be interpreted purely from the perspective of physical medicine; epiphenomenalism that other modes of explanation have their place, even if their results could in principle be deduced from the medical ones.[35] Galen did explicitly and with justice regard himself as a philosopher as well as a doctor, though this need not imply a distinction between the fields of competence of medicine and philosophy so much as the need to use the latter in studying the former. The title of another of Galen's treatises, or rather pair of treatises, *On the Diagnosis and On the*

Cure of the Passions and of the *Errors of the Soul*, certainly *sounds* medical enough; but it is the title and what it implies about approaching moral questions that are reminiscent of medicine, rather than the actual prescriptions in the work.[36]

Moreover, even in *That the Powers of the Soul Follow the Temperament of the Body* Galen does not suggest that medicine lays down the standards of moral behaviour, only that it can help us reach them by improving our diet.[37] And he could not consistently be claiming that what we eat determines every aspect of our behaviour – embracing, in other words, what we might term 'universal dietary determinism' – unless he accepted both that what he himself had eaten was the *sole* cause of his exhorting others to a better diet, and also that what they had previously eaten would determine whether they heeded his call or not. More plausible is Hankinson's view that for Galen the factors that lead to a certain sort of behaviour, including the decision to seek moral improvement, will include external factors, such as the people one associates with; the physiological conditions are only *necessary* conditions.[38] This, as is clear from Hankinson's discussion, involves no compromise either of Galen's materialism or of his view that punishment and reward do not depend on our being the cause of our actions in some strong sense; we may be as little responsible for our moral progress or lack of it as for the initial constitution we were born with, but we are justly punished or rewarded for the consequences of both. (Hankinson further argues that feedback must operate in the reverse direction too, so that moral decisions can alter physiological states indirectly [most obviously, by altering our diet] and perhaps also directly.)[39]

Galen's discussion provides us with a striking example from the ancient world of science, in our sense of the term, claiming greater competence concerning a particular area of human activity than some of his contemporaries would have allowed to it. But, despite Galen's concern to advance the claims of his medical art, he stops short of claiming that medicine has complete competence over human conduct and excludes all other considerations. More generally, it does not seem that we find in Greco-Roman antiquity attempts to eliminate all other approaches than that of natural science to the explanation of human activity and values. The possible exception is in the extent of Manilius' claims for astrology; and it is a typical one for the ancient world.

Notes

Most of this chapter is based on material presented at a seminar of the London Philosophy Programme in November 1996; it also incorporates parts of a paper on Popular Science and Natural Philosophy presented at a conference on ancient science, *Science Matters*, organised at the University of Liverpool by C. J. Tuplin and N. Fox in July 1996, some of the other papers from which are to be published in a volume forthcoming with Oxford University Press. I am grateful to all those who participated in the discussion on both occasions. In considering Galen's *That the Powers of the Soul Follow the Temperament of the Body* I have also greatly benefited from discussion at the 1997 Erasmus Summer School at Pontignano, Siena, organised by Antonina Alberti.

1 I will henceforth, for brevity's sake, use the term 'antiquity' to refer to the ancient Greco-Roman world specifically. Henri and H. A. Frankfort have suggested that a concern for consistent general explanations is one of the features that distinguishes philosophy from myth and justifies the claim that philosophy began with the Presocratics: see their contributions in Frankfort *et al*. 1949: 28–9, 253–4.

2 Philoponus, *On the Creation of the Universe* (*De Opificio Mundi*) 3.4, p.117.15 Reichardt (Leipzig: Teubner, 1897).

3 *SVF* (= von Arnim 1903–5) 2.939–44; Long and Sedley 1987, 55OP, 42E. Denyer 1985 indeed argues that observed correlations, though useful, are not essential to the Stoic account of divination which he interprets rather in terms of communications from the gods to us which might be one-off communications (5); thus, more generally, he warns against over-assimilation of Stoic divination to a scientific model.

4 It is for similar reasons that I do not here consider the case of neo-Pythagoreanism in later antiquity. It does not seem to me that the aspects of this which involve universal claims have much to do with its truly mathematical achievements. But I am conscious that I may be imposing anachronistic standards here. I am grateful to Paul Foulkes for raising this point.

5 On which see especially Lloyd 1979, chapter 1; van der Eijk 1990.

6 Aristophanes, *Clouds* 380: 'I did not realise that Zeus does not exist, but Vortex now rules instead of him'.

7 Cf. especially Galen, *On the Doctrines of Hippocrates and Plato*, 5.1, 5.4–5 (de Lacy 1978).

8 Aristotle, *On the Soul* 1.1 403a24ff. Tensions do emerge in the case of thought, which Aristotle at least seems to suggest may not be dependent on bodily organs; for a detailed account of passages in Aristotle's work which nevertheless suggest bodily influence even here, cf. van der Eijk 1997.

9 On the question of metaphysical priority as between form (including soul) and matter in various Peripatetic writers from Aristotle and Dicaearchus onwards see especially Caston 1997.

10 The metaphor of levels is explicitly used by Epicurus' follower Lucretius: 3.273–5.

11 On the specific question of human choice, however, Sedley seems to me to go too far in arguing that our volitions can actually *cause* the atoms of which we are made to swerve from the path they would otherwise follow; for this seems to be in conflict with the fundamental assumptions of the system. Cf. Sharples 1991–3: 176–82.

12 Plutarch, *On Stoic Self-Contradictions* 1045b. On the identity of Chrysippus' target in this passage see most recently Boys-Stones 1996.

13 And not only of ancient culture; John Donne, for example, adopts a similar approach. Whether it is actually our own age that is the exception, and if so what the reasons for this might be, are questions too large to go into here.

14 As is pointed out by Mette 1936: 12. (Again, I am grateful to Dr Broggiato for

this reference.) The imposition of modern habits of thought on the ancients remains all too easy, however. Wright 1995: 38 says, quoting the Homeric passage (rather than Crates himself), that 'the distances involved were first *calculated* by *timing* free-falling objects' (my italics). From the context Wright's intention is clearly only to say, as she does say of Lucretius on the next page, that the poet's intention was 'to help the reader to try to comprehend' the distance involved; but in a modern context, and in a series aimed at least in part at those who are not specialists in classical antiquity, misunderstanding seems a danger here.

15 Cf., for example, Seneca, *Moral Letters* 122.7–8, castigating the unnaturalness of forcing hothouse plants, and elaborating on the standard Roman criticism (cf. e.g. Horace, *Odes* 3.1.33, 3.24.3) of rich men who pervert nature by laying the foundations of their houses in the sea.

16 Hine 1981, 439, notes that Seneca draws on themes of consolation literature here.

17 This collection of sermon-illustrations from natural history, *Lumen Animae,* exists in three versions, of which only the second, *Lumen Animae B*, has ever been printed, four times before 1500 and never since. See Rouse and Rouse 1971; Thorndike 1934: 548–51; Dodwell 1961: xliv–lii; Sharples 1995: 22–3.

18 Hijmans 1991: 342 argues that *Ep: Mor*. 64.7–8 implies that Seneca transmitted and amplified an inheritance rather than contributing doctrines of his own. But the passage may be more ambiguous. Seneca begins with the image of an inheritance which one tries to increase before passing it on, and then suggests that even if the discoveries have been made one can contribute by finding practical applications for them. But it is not clear that the latter is the *only* role he allows for himself; and if it is not, the implication would seem to be that he is claiming – or at least not ruling out – discoveries of his own.

19 Alexander of Aphrodisias, *On Fate* 6 171.7–11 ≈ *De Anima Libri Mantissa* 186.8–9; for the Stoic view, *SVF* 2.1210, and cf. Tacitus, *Annals* 6.22.6: 'some things turn out otherwise than was said because of the deceitfulness of those who speak of things which they do not know. This damages the credibility of an art of which both ancient times and our own have given clear proofs.'

20 Alexander of Aphrodisias, *Quaestio* 2.16; Galen, *On the Method of Healing* 3.7, vol.10. 206.5–207.1 Kühn. Cf. also [Galen], *On the Best Sect, to Thrasybulus* 4, vol.1 114.13–115.4 Kühn. Ierodiakonou 1995, 481–2 and nn.25–6.

21 I say 'eliminates *moral* responsibility' because responsibility remains relevant only in the distinct sense of identifying the person who is principally involved in something's coming about and so is the person who is to be punished or rewarded for it if anyone is. Cf. Hankinson 1993, 216–17.

22 I am grateful to Andrew Gregory for raising this question.

23 On the other hand Hankinson 1993: 219–20 n.102, stresses the Platonic antecedents, especially in the *Laws*, of Galen's treatment (below). See Mackenzie 1981.

24 Notably in the discussion reported by Gellius, *Attic Nights* 7.2 = *SVF* 2.1000 = Long and Sedley 1987, 62D.

25 Lloyd's interpretation of this passage is similarly criticised by Hankinson 1993: 218 n.99.

26 The words 'by them' were deleted by Müller as a gloss meaning 'by the Greeks'; but Donini 1974: 142 n.29 argues that they should be retained and understood as 'by Platonists'.

27 'Education' is how Jowett, for example, translated the passage; the tendentiousness of Galen's interpretation is noted by Lloyd 1988: 20–2.

28 Lloyd 1988 23 points out that Galen tendentiously ignores other factors mentioned by Plato in the context; Plato refers to education (747ac) and malign supernatural influences (747e), as well as climate and water and products of the soil which might be thought to act on the body.

29 Sorabji 1993: 182 has however also claimed that Porphyry's restriction of vegetarianism to the intellectual elite is inconsistent with his general argument, and a concession to established political and religious norms.

30 64.19ff. Müller = 805 Kühn. Galen describes them as people who 'call themselves Platonists'. But such a position would seem to be in accord with Plato's *Phaedo* too.

31 If indeed we regard the science of diet as medicine; but Galen draws no distinction here.

32 38.19–39.6 Müller = 776 Kühn. Cf. Ballester 1988: 119, 135–7 and in general Nutton 1984: 320–3.

33 44.12–20 Müller = 782–3 Kühn. Andronicus' name, indeed, is only conjecturally restored in the text here.

34 Caston 1997. I am grateful to Richard Sorabji for drawing the question to my attention in the present context. Ballester 1988: 119 argues that 'Galen was not a radical materialist.'

35 Lloyd 1988 notes the vagueness in Galen's central thesis – just how much is implied by the claim that the character of the mind '*follows*' the temperament of the body? (33, 36) – and suggests that the vagueness is a deliberate attempt by Galen to make his case more plausible. Cf. Ballester 1988: 128–9. Lloyd also notes (35–6, 39) the emphasis in Galen's *On Temperaments* on the difficulty of analysing bodily temperaments and the effect of foodstuffs on them. Cf. Hankinson 1993: 186–7, 221; Singer 1997: xxxvi–xxxix.

36 Donini 1974: 132–3. Cf. Ballester 1988: 141–6 on the therapeutic use of reason.

37 I am grateful to Jo Wolff for this point.

38 Hankinson 1993: 222. Lloyd 1988: 41–2, suggests that Galen overstates his case in his concern to champion the cause of medicine against philosophers who would discount dependence of the mental on the physical altogether.

39 Hankinson 1993: 221, comparing modern discoveries concerning the effect of mental attitudes on the immune system. Cf. also Ballester 1988: 150–2.

Bibliography

Arnim, H. von (1903–5) *Stoicorum Veterum Fragmenta* (SVF), Leipzig: Teubner.

Bailey, C. (1947) *Lucretius*: *De Rerum Natura*, Oxford: Clarendon Press.

Ballester, L. G. (1988) 'Soul and body, disease of the soul and disease of the body in Galen's medical thought,' in Paola Manuli and Mario Vegetti (eds), *Le opere psicologiche di Galeno, Atti del terzo colloquio Galenico internazionale, Pavia, 10-12 September 1986*, Naples: Bibliopolis, 117–52.

Boys-Stones, G. (1996) '*The epeleustikê dunamis* in Aristo's psychology of action', *Phronesis* 41, 75–94.

Brunschwig, J. (1986) 'The cradle argument in Epicureanism and Stoicism', in M. Schofield and G. Striker (eds.), *The Norms of Nature: Studies in Hellenistic*

Ethics, Cambridge/Paris, 113–44.

Caston, V. (1997) 'Epiphenomenalisms: ancient and modern,' *Philosophical Review* 106, 309–63.

de Lacy, P. (1978) *Galen: De Placitis Hippocratis et Platonis* I-V (*Corpus Medicorum Graecorum* 5.4.1.2), Berlin: Akademie-Verlag.

Denyer, Nicholas (1985) 'The case against divination: an examination of Cicero's *De Divinatione*', *Proc. Cambridge Philological Soc.* n.s. 31, 1–10.

Dodds, E. R. (1951) *The Greeks and the Irrational*, Berkeley: University of California Press.

Dodwell, C. R. (1961) *Theophilus: The Various Arts*, London: Nelson.

Donini, P. L. (1974) *Tre Studi sull'Aristotelismo nel II secolo d.C*, Turin: Paravia.

Eijk, Ph. J. van der (1990) 'The theology of the Hippocratic treatise *On the Sacred Disease*,' *Apeiron* 23: 88–119.

Eijk, Ph. J. van der (1997) 'The matter of mind: Aristotle on the biology of 'psychic' processes and the bodily aspects of thinking', in W. Kullmann and S. Föllinger (eds), *Aristotelische Biologie: Intentionen, Methoden, Ergebnisse, Akten des Symposions über Aristoteles' Biologie vom 24.-28. Juli 1995 in der Werner-Reimers-Stiftung in Bad Homburg*, Stuttgart: Franz Steiner, 231–58

Fazzo, S. (1988) 'Alessandro di Afrodisia e Tolomeo: Aristotelismo e astrologia fra il II e il III secolo d.C,' *Rivista di storia di filosofia* 4, 627–49.

Frankfort, H. *et al.* (1949) *Before Philosophy*, Harmondsworth: Penguin (previously published as *The Intellectual Adventure of Ancient Man*, Chicago: Chicago University Press, 1946).

Gross, Nikolaus (1989) *Senecas Naturales Quaestiones: Komposition, naturphilosophische Aussagen und ihre Quellen*, Stuttgart: Franz Steiner.

Guthrie, W. K. C. (1965) *History of Greek Philosophy*, vol. 2. Cambridge: Cambridge University Press.

Hankinson, J. (1993) 'Actions and passions: affection, emotion, and moral self-management in Galen's philosophical psychology', in J. Brunschwig and M. C. Nussbaum (eds), *Passions and Perceptions*, Cambridge: Cambridge University Press, 184–222.

Hijmans, B. L. (1991) Discussion of O. Gigon, 'Seneca's *Naturales Quaestiones*', in *Entretiens Hardt 36, Sénèque et la prose latine*, Geneva: Vandoeuvres, 313–46.

Hine, H. M. (1981) *An Edition with Commentary on Seneca*, Natural Questions, *Book Two*, New York: Arno Press.

Ierodiakonou, K. (1995) 'Medicine as a stochastic art,' in Ph. J. van der Eijk, H. F. J. Horstmanshoff and P. H. Schrijvers (eds), *Ancient Medicine in its Socio-Cultural Context*, Amsterdam: Rodopi (*Clio Medica* 18) vol. 2: 473–85.

Jaeger, W. (1945) *Paideia* vol. 3, Oxford: Clarendon Press.

Jori, A. (1995) 'Le *pepaideumenos* et la médicine', in Ph. J. van der Eijk, H. F. J. Horstmanshoff and P. H. Schrijvers (eds), *Ancient Medicine in its Socio-Cultural Context*, Amsterdam: Rodopi (*Clio Medica* 18) vol. 2: 411–23.

Lloyd, G. E. R. (1979) *Magic, Reason and Experience*, Cambridge: Cambridge University Press.

Lloyd, G. E. R. (1988) 'Scholarship, authority and argument in Galen's *Quod animi mores*', in Paola Manuli and Mario Vegetti (eds), *Le opere psicologiche di Galeno, Atti del terzo colloquio Galenico internazionale, Pavia, 10–12 September 1986*, Naples: Bibliopolis, 11–42.

Long, A. A. and Sedley, D. N. (1987) *The Hellenistic Philosophers*, Cambridge: Cambridge University Press.

Mackenzie, M. M. (1981) *Plato's Theory of Punishment*, Berkeley: University of California Press.

Mette, H. (1936) *Sphairopoiia*, Munich: Beck.

Nutton, V. (1984) 'Galen in the eyes of his contemporaries,' *Bulletin of the History of Medicine* 58, 315–24.

Rouse, M. A. and R. H. Rouse (1971) 'The texts called *Lumen Animae*,' *Archivium Fratrum Predicatorum* 41, 5–113.

Sharples, R. W. (1983) *Alexander of Aphrodisias* On Fate, London: Duckworth.

Sharples, R. W. (1991–3) 'Epicurus, Carneades and the atomic swerve,' *Bulletin of the Institute of Classical Studies* 38, 174–90.

Sharples, R. W. (1995) *Theophrastus of Eresus: Sources for his Life, Writings, Thought and Influence, Commentary volume 5, Sources on Biology*, Leiden: Brill.

Singer, P. N. (1997) *Galen: Selected Works*, Oxford: Oxford University Press.

Sorabji, R. (1980) *Necessity, Cause and Blame*, London: Duckworth.

Sorabji, R. (1993) *Animal Minds and Human Morals*, London: Duckworth.

Theiler, W. (1946) 'Tacitus und die antike Schicksalslehre,' in *Phyllobolia für P. von der Mühll*, Basel, 35–90.

Thorndike, L. (1934) *History of Magic and Experimental Science* vol. 3, New York: Columbia University Press.

Wright, M. R. (1995) *Cosmology in Antiquity*, London: Routledge.

3 Theology, philosophy, and 'science' in the thirteenth century

The case of Albert the Great

M. W. F. Stone

Medieval 'science'[1] was but a part of a rich, complex and comprehensive intellectual outlook on the natural world and its place in the heavens.[2] That outlook was originally defined by Aristotle and other ancient philosophers, but enlarged in some cases and restricted in others by insights and ideas derived from the three monotheistic religious traditions of the West: Judaism, Christianity and Islam. The factors that framed this outlook came to be operative at different times and places, and they influenced individual thinkers in various ways. As a consequence, there was never, at any period of the Middle Ages, a uniform philosophical setting from which scientific thought, as we now know it, emerged.[3] Rather, medieval philosophical thought, itself neither authoritarian nor beholden to a monolithic theological agenda,[4] underwent an evolution that can be articulated into many movements and schools spanning recognisable chronological periods.[5]

Within the development of medieval philosophical thought, the period from the death of Pope Gregory IX in 1241 to the infamous condemnations of Etienne Tempier, Bishop of Paris in 1277, provides an accessible setting for a philosophical understanding of medieval science. Explaining such a setting will be the burden of this chapter. However, in order that the scope and point of high medieval science can be made tractable to the reader, I have decided to select one example from this period of study. To this end, I shall focus on an intellectual giant of the thirteenth century, Albert the Great (*Albertus Magnus*) (ca. 1200–1280)[6] and his influential discussion of the so-called 'hierarchy of the sciences'. In order that Albert's discussion of this important concept can be made coherent, I shall begin by saying something about the intellectual context of the early and mid-thirteenth century and how it came to fashion his very distinctive natural philosophy.

The fear and fascination of Aristotle

In the thirteenth century the scope of philosophy was quite broad, encompassing everything that can be known about the universe by the use of reason alone, unaided by any special revelation.[7] Theology or sacred doctrine (*sacra doctrina*), thus fell outside the remit of philosophy, and so did practical disciplines such as grammar, mechanics and medicine. Ethics, then as now, pertained to philosophical discourse, as did logic, natural philosophy and metaphysics, thereby reflecting a common division of philosophy that had come down to the Middle Ages from antiquity. Epistemology, especially as it is practised in contemporary English-speaking philosophy, did not yet exist, although the speculative issues connected with the study of human knowledge, its limits and objects did interest many thinkers of the period. Psychology, the study of the rational soul, was regarded as a branch of natural philosophy, as were all the disciplines that are now viewed as the natural sciences: astronomy, cosmology, chemistry, physics and biology.[8] Mathematics was seen as belonging to philosophy, broadly construed, even though there was little agreement on the manner in which mathematical reasoning was to be related to natural philosophy.[9]

Central to the study of these subjects were the works of Aristotle and his ancient commentators. Most of Aristotle's works and some of the commentaries on them – especially those of the eleventh-century Arab philosopher Avicenna (980–1037) – were made available to Western scholars in Latin translations by 1200. Very little is known about their early circulation or their pedagogical use, but they seem to have made an appearance at both the Universities of Paris and Oxford during the first decade and a half of the thirteenth century.[10] The explanatory power of the Aristotelian system and its promise of untold future benefits in all facets of learning proved very seductive to this generation of medieval thinkers, but it bestowed its benefits at a certain price. For Aristotelian philosophy inevitably courted controversy either by virtue of its intellectual challenge to specific points of Christian doctrine,[11] or else in its attempt to supplant the blend of Neoplatonic philosophy and Christian theology that had been dominant in the West from the time of Augustine (354–430).[12] While it had long been thought legitimate for theologians in the Christian tradition to make use of Aristotle and other ancient philosophers as they saw fit,[13] the problem in the mid-thirteenth century arose from the fact that Aristotelian philosophy did not fill a conceptual vacuum but instead invaded occupied territory. The temerity of this intellectual occupation was to lead to many disputes which called forth for resolution. As we shall see, the natural philosophy of Albert the Great was, in part, proposed as a means of resolving the quarrel between the new learning and accepted dogmas. In another respect, it was put forward as a speculative account of the working of the world of nature.

It was at the University of Paris that Aristotle's work first encountered difficulty. Allegations were made that a form of pantheism (in this instance, the identification of God with the universe) was being taught by some masters of arts under the influence of the Stagirite. The outcome of these charges was a decree, issued by a council of bishops in Paris in 1210, forbidding the instruction of Aristotle's natural philosophy within the faculty of arts. The decree was renewed in 1215.[14] In 1231 Pope Gregory IX became directly involved in the dispute in the course of promulgating regulations governing the University of Paris. Gregory acknowledged the legitimacy of the ban of 1210, and renewed it, specifying that Aristotle's books on natural philosophy were not to be read in the faculty of arts until they had been examined and purged of all 'suspected error'.[15]

Other documents address the fortunes of Aristotle's works at this time. They reveal that while the bans of 1210, 1215, and 1235 were moderately successful, they began to lose their effectiveness around 1240. One reason for this may have been Gregory's death in 1241, while another may have been the growing awareness among the Parisian masters that they were losing ground to their colleagues in Oxford, who had adopted a more liberal attitude to Aristotelianism.[16] Yet whatever the causes, Aristotle's works on natural philosophy seem to have become the subject of lectures and disputations in the Arts Faculties around 1240. This process continued well into the 1250s by which time Aristotelian natural philosophy had not only created a place for itself in the arts curriculum, but had become one of its principal features.

One of the first significant figures to comment on Aristotle at this time was Robert Grosseteste (ca. 1168–1253). Though not a Franciscan,[17] Grosseteste was the first lecturer in that order's *studium generale* at Oxford. His commentary on Aristotle's *Posterior Analytics*, written sometime in the 1220s, was one of the earliest efforts at this time to deal seriously with Aristotle's scientific method.[18] Grosseteste was also acquainted with Aristotle's *Physics, Metaphysics* and *Meteorology*, and biological works. Despite his familiarity with these texts, Grosseteste's intellectual formation was strongly shaped by Platonic and Neoplatonic influences, and also by some of the new works on mathematical science. Thus, in his works on physical science we find an uneasy juxtaposition of Aristotelian and non-Aristotelian elements. This can be seen in his account of cosmogony, which is set within a broadly Aristotelian framework but which attempts to reconcile Neoplatonic emanationism – the idea that the created universe emanated from the deity as light emanates from the sun – with the *ex nihilo* account of creation which is to be found in the opening chapters of the Book of Genesis.[19]

Important aspects of Grosseteste's programme were continued by Roger Bacon (ca. 1220–1292). The details of Bacon's education are unclear, but it is known that he studied at both Oxford and Paris, where he was one of the first to lecture on Aristotle's books on natural philosophy – *Metaphysics*, *Physics*, *De Sensu*, *De Generatione et Corruptione*, *De Anima*, and *De Caelo*. Later he joined the Franciscan order and spent the remainder of his life in study and writing. An important aspect of Bacon's writings is their polemical tone. Bacon was a committed 'evangelist' for the natural sciences and his writings reflect a systematic attempt to persuade the church authorities of the value of the new learning which he defined as a divine gift.[20] Among other things, Bacon argued that the new learning was capable of proving the articles of faith; that astronomy is essential for understanding the religious calendar; that experimental science can teach us how to prolong life; and even that the science of optics can enables us to create devices that will terrorise unbelievers and lead to their conversion![21] The natural sciences were thus justified by their religious utility; theology does not oppress science, it simply directs it to its proper end.[22]

Despite Bacon's enthusiasm, many members of his own Franciscan order developed a more cautious attitude to the new Aristotle. One of the most influential of this number was the Italian Bonaventure (ca. 1217–1274). Bonaventure studied both the liberal arts and theology at the University of Paris, then remained to teach theology, eventually resigning to become minister general of his order. While there is little doubt of Bonaventure's profound respect for Aristotle, his more general intellectual allegiance to the philosophical tradition of Augustine and Neoplatonism was to cause him to be much more theologically suspicious of Aristotelian learning than either Grosseteste or Bacon.

Bonaventure certainly agreed with Bacon on the applicability of the Augustinian formula that pagan philosophy was an instrument to be used for the benefit of Christian theology. But he was much more sceptical than Bacon of the utility of philosophy *per se* and more sharply aware of promoting it, particularly in its Aristotelian mode, among the intellectual community. The locus of Bonaventure's complaint resided in his account of the origins of human knowledge. Without the assistance of divine illumination, he thought, we would not be able to know anything. Thus Bonaventure was apt to keep Aristotelian philosophy on a very short lease and to abandon it at that point at which he believed it to be in conflict with revealed teaching.[23]

In the persons of Grosseteste and Bacon and in the later work of Bonaventure we can see several important tendencies of the thirteenth century: a growing knowledge of the Aristotelian corpus, a mixture of admiration and suspicion about its contents, and a tendency to read various Augustinian or Platonic ideas into Aristotelian texts. It was to be left to two great Dominican friars, Albert and his pupil Thomas Aquinas (1224–1274)[24] – whose achievements will not figure in the following pages – to adopt a more open attitude to Aristotle, an attitude that was to be one of the main defining features of high medieval science.

Albert's appropriation of Aristotle and his scientific achievements

Albert was the first to offer a comprehensive interpretation of Aristotle's philosophy in Western Christendom and on these grounds he is often credited as the effective founder of a 'Christian Aristotelianism'. This last term, however, should not be taken to mean that Albert ever achieved a form of philosophical purity; some of his early commentaries were devoted to Neoplatonic authors, and to the end of his life he retained a strong allegiance to many components of Neoplatonic philosophy.[25] Moreover, in common with other thirteenth-century philosophers he was always ready to correct or discard those Aristotelian doctrines that he considered false according to the requirements of Christianity.

None the less, Albert perceived the profound significance of Aristotelian philosophy and set out to interpret it for his fellow Dominicans.[26] In the prologue to his commentary on Aristotle's *Physics*, he explained:

> Our purpose . . . is to satisfy as far as we can those brethren of our order who for many years now have begged us to compose for them a book on physics in which they might find a complete exposition of natural science and from which they might be able to understand correctly the books of Aristotle.
>
> (*Physica* 1.1.1. (Borgnet 3:2a))

Albert responded to this pedagogical challenge not only with a *Physics* commentary but with commentaries on, or paraphrases of, all the available Aristotelian books. His purpose in doing so was to exhibit and make available the explanatory power of Aristotelian philosophy, which he regarded as a necessary preparation for theological studies. He had no intention of releasing Aristotelian philosophy from its subordinate role to Christian theology, but he did mean to give it substantially larger responsibilities.

One of the familiar features of Albert's work is the number of paraphrases he wrote on Aristotle's works. The paraphrases enact a single project, to make all areas of philosophy intelligible to the Latins (*nostra intentio est omnes dictas partes [scil. philosophiae] facere Latinis intelligibilis*) (*Physica* 1.1.1. (Borgnet 3:2a)). The chronology of the paraphrases has been fairly well worked out by James Weisheipl,[27] but their relation to the rest of Albert's corpus has not. At several points in the text Albert disclaims the teaching of the paraphrases. They are to be read he insists, not as expressions of his own mind, but as expressions of the sense (*intentio*) of Aristotle.[28] Yet the paraphrases range widely in their choice of topics and interlocutors, and Albert frequently interrupts the plain reading of Aristotle to indulge in a controversy or to offer erudition.

The contrast with other thirteenth century commentaries on Aristotle, such as those offered by Aquinas, is often quite striking. While Aquinas enters no explicit disclaimers, he restricts himself, for the most part, to an interpretation of the *intentio* Aristotelian text or least the Latin versions of Robert Grosseteste or William of Moerbeke.[29] What the Albertine passages have to say on the relations that exist between the sciences, for instance, is clearly intended to be more than a simple exposition of Aristotle. The paraphrases have as their aim the recasting of an idealised Aristotelian corpus into a comprehensive presentation of philosophy. Albert promises as much in the *Physics* commentary and then fulfils the promise in *De Intellectu*.

Armed with a thorough understanding of Aristotle, Albert set about applying the new scientific learning to a broad range of topics. Here, as elsewhere, his reliance upon Aristotle was by no means total. He drew upon Avicenna in psychology, the then available works of Plato, particularly the *Timeaus*, in cosmology, Euclid in geometry, as well as Galen (to a limited degree), Averröes, Constantine the African, and a host of other Greek and Latin authors in other scientific disciplines.[30] His chief concern in appropriating such sources was to bring them to bear on problems he had confronted when interpreting Aristotelian texts.[31]

By the standards of any age Albert's scientific achievements were considerable. He was an acute observer of plant and animal life. For example, he is well known as having corrected Avicenna's theory on the mating of partridges, and he was a frequent visitor to birds' nests in order to learn more about their patterns of behaviour.[32] Further to that he makes a genuine claim to be one of the first botanists of the medieval period.[33] His intellectual energy was boundless and his enthusiasm for the natural sciences quite genuine. Less than half of his extant corpus is devoted to works on scientific and what we would now refer to as 'pseudo-scientific' subjects. These treat subjects as diverse as: physics, astronomy, astrology, alchemy, mineralogy, physiology, medicine, natural history, psychology, logic and mathematics. The authority both conceptual and experimental with which Albert could address these disciplines, explains why even in his own day he was referred to as 'great' (*magnus*).[34] Yet despite Albert's considerable scientific prowess and successes, his enduring influence on medieval science and subsequent reflection upon it is to be observed in his remarks on the 'hierarchy of the sciences'. There, his scientific knowledge, philosophical insight and theological acumen are brought together in a synthesis which is one of the defining features of high medieval thought.

The hierarchy of the sciences

Albert's reflections on the topic of the hierarchy of the sciences are important for several reasons. As has been explained above, he was at once the boldest and most assiduous Aristotelian commentator of his time, a time by no means lacking in thoughtful commentators. Further to this, Albert was justly renowned among his contemporaries for having mastered so many sciences. If anyone had traversed the hierarchy of the sciences in all its details he certainly had. Finally, given Albert's understanding of his own role in terms of Dominican teacher of theology, he was required to address the role of revealed theology among the other *scientiae*. For these reasons alone, what Albert has to say on this topic is very relevant to our task of sketching the philosophical setting of high medieval science.

What did Albert understand by the 'hierarchy of the sciences'? By this phrase he would have understood a host of teachings and questions which had been handed down to his own century by various ancient and patristic texts. The topic of the hierarchy of the sciences can be said to begin with Plato and his discussion at *Republic* 6 and 7. There, the Platonic Socrates considers together the soul's powers, their objects, languages for describing them, their conversion by philosophy, and their perfection in the education of future rulers. The *Republic*, of course, was not available to medieval readers. They had to learn of the topic from Plato's successors.[35] Early medieval readers could discover one such succession in Boethius (c.480–c.526) on Aristotle, but others were available in Augustine's retelling of Stoic and Neoplatonic philosophy, in Seneca the Younger (c.4 BC–65 AD) and Cicero (106–43 BC), in the encyclopaedias of Martianus Capella (fifth century AD, dates unknown) and Isidore of Seville (before 534–636), and in the treatises of Pseudo-Dionysius (fl. 500).[36] By the thirteenth century, there were added to these texts not only Aristotle's several accounts of the hierarchy, but variations on them and on Plato by Alfarabi (875–950), Avicenna, and Averröes (1126–1198).[37]

The compound of Platonic and Neoplatonic teaching on the hierarchy of the sciences is passed down to an author like Albert in numerous and complicated ways. In the first place, it had been diluted by the Peripatetic tradition of late antiquity in so far as that tradition had succeeded in identifying certain issues for examination and had further provided a technical vocabulary for their discussion. Albert did not approach the topic unaided, as for many centuries before him Christian and Islamic authors had extended the hierarchy of the sciences to include theologies derived from revelation. Prior to Albert, then, the questions and issues surrounding the hierarchy of the sciences had become a firmly established topic for reflection.

Albert discusses the hierarchy of the sciences most directly and consecutively in his paraphrases of Aristotle. He also treats it in other passages, as say in those passages attached to the Dionysian corpus.[38] As a complete list of Albert's scattered remarks on the hierarchy of the sciences would be difficult to compile and to exploit intelligently, I have chosen instead to concentrate on three texts which together span more than two decades of Albert's thinking. I shall begin with the paraphrase of Aristotle's *Physics* (completed just before 1250), move next to the short treatise *De Intellectu* (completed sometime between 1254 and 1260), and conclude with the *Summa Theologiae* (finished after 1274). Going on from there, I shall argue that the three texts not only address different aspects of Albert's thinking on the relations that exist between 'science', theology and philosophy, but that they reveal him to have made genuine progress in the appreciation of these relations.

The Aristotelian paraphrases

Albert's paraphrase of Aristotle's *Physics* beings with a prologue that aims to make tractable the methods and procedures of Aristotelian science. It notes and explains Aristotle's division of theoretical knowledge into metaphysics, mathematics and physics. First in the order of nature is metaphysics or *theologia* (theology), the universal science that considers being (*ens*) qua being, so far as being is conceived as without motion or sensible matter. After this come mathematics, which considers its objects 'with motion and sensible matter according to [their] manner of existing (*esse*), but not according to the account (*ratio*) [to be given of them]' (*Physica* 1.1.1. (Borgnet 3: 2a)). Of the three, physics is the lowest science since its conceives its object with motion and sensible matter both according to their manner of existence and according to their *ratio*. By way of illustrating the objects appropriate to the three sciences, Albert cites substance for metaphysics, a Euclidean circle or line for mathematics, and an element, or something constituted by an element, for physics. He then seeks to extend his remarks on physics by making a further distinction between physical and logical definitions. Natural things, Albert says, must be defined in physics by reference to sensible matter and concrete subjects. Such definitions stand in contrast to those in the domain of logic which make reference to the common notions of genus and difference, where these are understood to be simple and universal. Being such, simple and universal definitions are not a part of physics (*Physica* 1.1.1. (Borgnet 3:3a)).

Having set down the basic scheme of the three sciences, Albert then seeks to augment his understanding of these divisions. He does this by assigning further content to the objects of the three sciences. Thus, metaphysics considers only the intelligible; mathematics the intelligible and the imaginable; while physics treats the intelligible, imaginable and sensible (*Physica* 1.1.1. (Borgnet 3:3a)). Each of the sciences corresponds Albert argues to an understanding of body. Metaphysics considers body simply and without qualification retracing it and eventually subsuming it within being. Mathematics treats 'intelligible matter' or 'imaginable quantity' according to the various figures that can be constructed according to its methods and procedures.[39] Physics treats natural body universally, while its branches consider one or other aspect of physical body. Such sub-divisions within the subject of physics reveal it to be a 'general science', that is, a group of particular sciences. Mathematics is also for Albert a general science. Only metaphysics, he thinks, is the 'universal science'. For that reason metaphysics cannot be sub-divided into different fields of inquiry.

Albert then proceeds to draw out those implications that the existence of the trichotomy of the sciences may or may not have for human learning. He begins by stating that the objects of metaphysics are the causes of the objects of mathematics and physics. Therefore, the principles of mathematics and physics can be extracted from those of metaphysics, for in that domain of inquiry they are tested and proved (*probata*) (*Physica* 1.1.1. (Borgnet 3:3b)). Albert holds that the relation between metaphysics and the other principal sciences can be made plain by reference to the model of whole and parts: metaphysics is the universal science, while mathematics and physics are more restricted modes of inquiry that treat only parts of being (*ens*) rather than being qua being. Only metaphysics can demonstrate principles properly by means of its universal grasp of being qua being. The logician can also construct some version of the other sciences since he has recourse to the common argumentative principles of all sciences. Here, Albert thinks, the discursive universality of logic resembles the genuine universality of metaphysics.[40] Logic can also instruct us that true science proceeds demonstratively from first principles to proximate principles. Any other mode of argument, Albert contends, will be 'topical', that is, dialectical or rhetorical (*Physica* 1.1.5. (Borgnet 3:10b)).

In itself, the abstract nature of the downward flow of causes from metaphysics to physics in the hierarchy of the sciences, may be too difficult to comprehend for human beings who study the relations that exist among the sciences. For this reason, Albert argues that any instruction concerning the hierarchy must commence with what is easiest and nearest to hand. Good teaching begins, he argues, from what can be grasped most immediately by sense, imagination and intellect. Moreover, any instruction in the lowest order of learning, physics, must begin with the acquisition of those principles that will help us to establish particular conclusions about the operations of the natural world. Hence Albert construes the ancient half-title of Aristotle's *Physics, ex auditu*, to mean that the principles and arguments given in the book are instantiated by hearing rather than by demonstration (*Physica* 1.1.5. (Borgnet 3:10b)).

The requirement that one commence one's instruction of the sciences with what is already known leads Albert to entertain a number of interesting inversions of the hierarchy. In turn, this leads him to consider more complex questions about the independence of the sciences. He reads in Aristotle's book that every science is concerned with principles (*principia*, cf. Aristotle's ἀρχαί), causes (*causae*, cf. αἴτια) and elements (*elementa*, cf. στοιχεῖα); see *Physics* 1.1. 184a1–2. Albert then argues that is only in the field of physics that one can learn *principia, causae*, and *elementa* as such. While metaphysics considers every cause, it treats form and matter not as constitutive of things, but rather reducible to the understanding of substance (*Physica* 1.1.5. (Borgnet 3:1la)). This last remark invites the following question: 'How can it be that something necessary to all scientific demonstration is to be learned fully only in the field of physics?' Albert answers this query by stating that *principia* and *causae* are sufficiently known in each science in order for them to be tractable objects of learning. He then offers a further argument for this view that is based on an distinction between prior axioms (*dignitates*) and proximate axioms. Albert writes:

> Sometimes demonstration proceeds from what are first and true, and these are principles (*principia*). Sometimes however [it proceeds] from what are taken on faith (*fides*) from what are first and true, which [things] however are not proximate. And sometimes [it proceeds] by what are last and essential (*essentialia*). Therefore what are first and true are the principles. And what are from the first and true, and accepted further in order to infer other things, are the causes: since propositions are the cause of the conclusion. And the elements are proximate, [and] they are required to prove nothing.
>
> (*Physica* 1.1.5. (Borgnet 3:11a–12b))

For Albert, then, not every science must proceed from *principia, causae* and *elementa* in the same way. That said, the above passage does appear to suggest that the higher sciences such as metaphysics and mathematics can appropriate the pedagogy of causes from physics on 'faith', that is, as something already known.

Going on from there, Albert notes that there is a difference in physics between the order of nature and the order of our learning. Our learning will always begin with a confused and partial apprehension of the obvious, or 'most common things'. This less than satisfactory initial apprehension of the commonplaces of nature must then be analysed or divided according to the methods and procedures of physics, in order that we can discover the special efficient causes and the particular elements of natural objects (*Physica* 1.1.6. (Borgnet 3: 12b)). In others words, the methods of physics enable us to ameliorate our ordinary knowledge of the natural world. Albert illustrates this last point with two examples. We know, he says, the whole circle that is to be defined before we know the components of its Euclidean definition. Likewise, a child begins by calling all men 'father' and women 'mother', and only then picks out the particular man who is his father and the particular woman who is his mother.

For Albert, human learning engages with the physical world in much the same way. This, he argues, is sufficient to distinguish the manner in which humans derive knowledge from physics, from the type of knowledge they derive from the demonstrative deductions of metaphysics and mathematics. Albert distinguishes three culminative stages of sense awareness: the apprehension (*acceptio*) of a particular sense (*acceptio secundum sensum particularem tantum*); the apprehension of the common power (*acceptio secundum communem simul et particularem*); and the apprehension of what he determines as 'a certain cognition of confused reason' (*acceptio secundum sensum particularem et communem et aliquam cognitionem confusae rationis in sensu vel cognitionis quae loco rationis est*) (*Physica* 1.1.6. (Borgnet 3:14b)). With regard to the processes of human learning Albert's threefold distinction can be explained in the following way. The human learner starts as it were with the highest genus, substance, and then works downwards until the indefiniteness is resolved into a very specific individual (*individuum signatum*). Here Albert reiterates Avicenna's example of recognising someone at a distance: first, one sees something (a substance), then one recognises that something is moving (an animal), then something erect (a human being), and finally something with particular properties (Socrates). For Albert the recognitional process described in Avicenna's example is typical of our actual psychology, and one which further reflects the development of our brain physiology. Moreover, and significantly, this process of composition, which moves from an indistinct simple to a distinct composite object, is proper to physics and to no other science. Thus, it is in the field of physics that we learn about the workings of nature.

Having briefly surveyed the course of Albert's argument in the prologue to Aristotelian science, we can note that his account of the hierarchy of the sciences and their explicit and implicit relations appears to have pulled in several directions. On the one hand, it is clear that the order of abstraction is the reverse of the order of human discovery. Albert holds that we discover things about the world by moving from vague abstraction to definite particulars. Moreover, a similar reversal seems to occur when we consider the *principia*, for the Aristotelian threefold division of the sciences suggests that the principles of every science will be secure only at the very summit of the hierarchy, in metaphysics. On the other hand, Albert argues that it is only in physics that the central terms of knowledge are to be learned. Again, there appears to be an inversion in the hierarchy in the curious relation that might be said to hold between metaphysics and logic. Logic provides what might be said to be an abstract anticipation of the axioms of metaphysics. Yet logic is preliminary even to physics and so stands at the bottom rung of the hierarchy, sandwiched between the liberal arts and philosophy.

To be fair to Albert he does attempt to tackle the problems caused by his inversions and reversals of the hierarchy at other places in the Aristotelian paraphrases. In his paraphrase of the *Categories*, for example, he considers at length the issue whether logic is a distinct discipline or whether it is a part of philosophy (*Praedicabilia* 1.1.1–2). Such attempts aside, it ought to be noted that the very fact that Albert has sought to reverse and inverse the traditional Aristotelian hierarchy is significant, for it provides good evidence of the uneasy alliance of certain Neoplatonic and Aristotelian components that are continually present in his thought. The Neoplatonic influence in Albert's thinking about the hierarchy of the sciences is centred around a narrative of human learning in which one progressively ascends the hierarchy thereby improving the scope and the content of one's knowledge. This stands in contrast to the Aristotelian threefold division of the sciences which begins as a division of objects and then proceeds to match them with accomplished bodies of ideal teaching, that is, demonstrations. Since Aristotle was notoriously silent in regard to the discovery of principles it is not unsurprising that Albert seeks to fill in this gap by bringing to bear a Neoplatonic account of the process of discovery.[41] It is significant, then, that whenever he tries to explain and describe human learning in his Aristotelian paraphrases, Albert can be said to exacerbate the tensions that inhere within the basic Aristotelian account, for what interests him is not so much a simple taxonomy of human knowledge, but rather a rich account of the mind's ascent along the hierarchy. This provides good evidence for the view that his interest in the topic of the hierarchy of the sciences, and by implication the orders of human knowledge, is as much a product of his allegiance to Neoplatonic metaphysics as it is a sign of his enthusiasm for Aristotelian natural philosophy.

De Intellectu

The most striking of Albert's three main treatments of the hierarchy of sciences is in *De Intellectu et Intelligibili*. He composed it to fill a gap in the *Parva naturaliter*, these being Aristotle's short works on biological topics. The gaps falls textually between *On Sense and What is Sensed* and *On Sleep and Wakefulness*. Albert argues that there can be no discussion of sleep until there has been a further study of the intellect beyond what has been said in Aristotle's *De Anima* Book III.[42] So he interposes a short treatise in three parts: on the nature of the intellect as such (1.1), on the intelligible as it belongs to intellect (1.2), and on the unity and diversity of the intellect with regard to the intelligibles (1.3). To this treatise Albert then adds a second book on the natural completion or perfection of the intellectual soul. While the topic of the hierarchy of sciences figures at several points in the first book, the narrative of ascent preoccupies the second.

The Aristotelian trichotomy of sciences appears first towards the end of

Book 1, in the exposition of the kinds of intelligibles.[43] Albert wants to compare the objects of the three sciences with the limited power of human intellection. Metaphysical or theological objects exceed this power, mathematical objects are proportioned to it, while physical objects fall below it in regard to certainty and firmness because of privation, matter and motion (*De Intellectu* 1.3.2 (Borgnet 9:500a)). This gradation within the powers of understanding is reflected in the sequence of demonstration itself. The principles of demonstration 'have much of the light and form of the intellect', but conclusions have less of this light and have it only as an effect. Thus the habitual possession of principles is given one name, '*intellectus*', while the possession of conclusions takes another, '*scientia*'. Reasoning (*ratiocinatio*) is the path (*discursus*) of light from principles to conclusions.

Albert appends to this account a gradation of intellects, that is, a ranking of different intellectual powers and states (*De Intellectu* 1.3.3. (Borgnet 9:501a)).[44] His remarks begin with a discussion of the possible and active intellects described by Aristotle at *De Anima* Book III. Next comes the formal intellect, which arises when intellectual light produces a form in the soul. Formal intellect is divided first into practical and speculative components. It is next divided, more importantly, into simple and composite. Simple formal intellect is 'an intelligence of non-complex things.' Composite formal intellect is 'an intelligence of complex things' compounded by enunciation, syllogism or another form of argument. Composite formal intellect is further divided into an inborn intellect of principles and an accomplished intellect (*intellectus adeptus*) acquired by discovery or study.

The schema of intellectual powers and states foreshadows the structure of *De Intellectu Book* 2. Albert expands the schema there by inserting the 'effected' or actualised intellect between the intellect of principles and the accomplished intellect, then by adding an assimilative intellect. The effected intellect (*intellectus in effectu*) is intellect actualised by an abstracted form that has been illuminated by the agent intellect (*De Intellectu* 2.6. (Borgnet 9:512a)). The assimilative or assimilating intellect 'is that in which a human being, so far as is possible or permitted to him, rises proportionately to the divine intellect' (*De Intellectu* 2.9 (Borgnet 9:516a)). Thus the complete hierarchy of intellects is given by Albert as possible, agent and formal (or of principles), effected, accomplished, and assimilative or divine (*De Intellectu* 2.9 (Borgnet 9:517a–517b)).

This hierarchy of intellects was known to Albert from several sources, none of which he ever appropriated in an unmodified form. Let me mention two. The first is Alfarabi's brief treatise also entitled *De Intellectu*. This work describes a sequence of five intellects: in potency, in effect, accomplished, agent, and divine.[45] The agent intellect appears at the end of the list because it is, for Alfarabi, a separate form, the cause of other intellects, and the giver of forms to bodies. It stands one step below the divine. Albert could not accept such a description, and so he changes both the agent intellect's position in the hierarchy and its attributes. Consider, as a second source, Avicenna's *De Anima*, 5.6. The sequence of intellects is less orderly here, but one does read of a material intellect, an intellect in effect, an agent intellect, and an accomplished intellect.[46] To these, Avicenna adds what he calls the 'holy intellect' which is the pre-eminent virtue of prophecy (Avicenna, *Liber de Anima* 5.6 (Van Riet 151. 84–5 and 153. 15–7). Albert too will speak of a holy mind, but will do so as something known to philosophy.

Whatever its sources, what is to be learned from this hierarchy? It serves to show, I think, several things. First, and most simply, Albert can diagnose by it what is needed for different kinds of minds (*De Intellectu* 1.3.3 (Borgnet 9:501b)). Some minds are naturally more attached to imagination and sense; they are incapable of learning except by sensible examples. They cannot attain to metaphysics. A second or middle mind is helped towards understanding by prophetic or divine teaching. A third kind of mind is born to understand things by itself or with only slight teaching. This is the philosophical mind.

The hierarchy permits Albert to explain something of how human learning is fulfilled in a glimpse of the divine. While Albert emphasises at the beginning of *De Intellectu* Book II that he intends to speak about the degrees by which the intellect rises to completion, he is not concerned here with the intellect's ultimate happiness (*De Intellectu* 2.1 (Borgnet 9:503a–b)). The whole discussion takes place within the ambit of philosophy not of Christian theology. It describes the highest condition of the assimilative intellect as a vision of the divine and in terms of a rational working that is perfected by holiness.

Albert characterises the assimilative intellect as an ascent to the highest cause of intelligences and intelligibility. The ascent is possible because the divine light has made itself manifest at four lower levels of apprehension (*De Intellectu* 2.9 (Borgnet 9:516b)). It reaches down first to the essential differences of each thing, making them manifest to lower powers. It is, second, the light in which intelligible forms have the being of abstraction or separation for the possible intellect, which is its image (*imago*). The third manifestation is that in which there appear universally in the agent intellect truths that come from the light of the inner microcosm (*minor mundus*). The divine light is joined to the agent intellect not as light to darkness or privation or potency, but as one light to another. The fourth illumination, the last, is what is manifest in the application of divine light to the intelligences that move the celestial spheres. It is here that Albert gives full range to his Neoplatonist sympathies.

The account of the divine mind that discloses itself along the steps of the hierarchy makes clear one thing that was not present in the Aristotelian trichotomy. Ascent up the levels of understanding is realised in progress in self-knowledge. Albert has already stressed that the human mind is an image of the divine mind. For him our mind's work is to find itself amid the distraction of bodies (*De Intellectu* 2.8 (Borgnet 9:515a) with references to Plato and Alfarabi). Albert now prescribes four stages for the mind's self-discovery. First, what must be described are the objects to be thought and their effects on the mind thinking. Then the mind is to be freed from flesh, from time and extension, and from matter so that it turns to the divine and so becomes more itself. Here Albert's language assumes an unusual degree of colour as he quotes passages from Apuleius and Pseudo-Dionysius (*De Intellectu* 2.10; for Apuleius 2.10 and Pseudo-Dionysius 2.11 (Borgnet 9:518b and 519b)). Above all, however, he insists that ascent along the hierarchy of intellects is the human soul coming into knowledge of itself: of its descent from the divine, of its role as pedagogue in rationalising the corporeal world, and of its own immortality (*De Intellectu* 2.12 (Borgnet 9:520b)).

The mystagogical conclusion of *De Intellectu* ought not to dispel doubts that Albert has not been true to his claim that he would only speak of philosophical matters. It is certainly true that Albert has philosophical sources in which mystical contemplation is discussed. But it is equally true that he cannot appropriate these sources as they stand. For a Christian *theologus*, the agency of ascent to the divine is not simply a diffuse illumination. It is a gift of grace announced and accomplished through a specific revelation. So what Albert has gained in providing a narrative of ascent, he must now secure by naming the power that makes ascent possible.

Summa Theologiae

The *Summa Theologiae* was certainly the last of Albert's major works to be finished, and the main part of it was probably composed after the Aristotelian paraphrases.[47] It begins, by well established tradition, with a prologue and disputed questions on theology as a science. The prologue follows one of the fixed patterns for an *accessus* or introduction: a Scriptural verse (*prothema*) is dissected phrase by phrase to uncover the outline of a doctrine about revealed theology. The disputed questions raised by Albert after the prologue are also traditional, though his way of arranging them and subordinating them seems original, at least by the standards of the second half of the thirteenth century.[48] However standard the form of his *Summa*, Albert's teaching in it on the relation of theology to the other sciences provides a necessary supplement to his remarks on the teaching of the Aristotelian paraphrases and *De Intellectu*.

The *prothema* for Albert's prologue is Psalm 138, verse 6.[49] He reads it as claiming that theology is higher than all the other sciences in six ways: in honour or nobility, in origin, in trustworthiness, in applicability, in demonstrative force, and in infinity of its object (*Summa Theologiae* 1. prologue (IAM 34: 1.5–17)). Each of the six points is explicated by reference to philosophical teaching and to Scripture. So, for the first point, Albert reminds the reader of Aristotle's remarks in *De Anima* about the wondrousness of knowing about the soul. Theology, Albert contends, is more wonderful, more honourable and more noble. On the second point, again, Albert juxtaposes Alfarabi with Scripture and Augustine to argue that only theology 'seals' the soul with divinity itself.

Significantly, Albert uses the third heading to assert that theology is the only true science (*planum est hanc vel solam vel praecipuam esse scientiam*) (*Summa Theologiae* 1. prologue (IAM 34: 2.25–6)). He argues that even if the other sciences deal with immobile intelligibilities, they still learn of them by reflecting on moving creatures. Only theology is grounded in the eternal *rationes*, completely removed from motion. Thus only theology completely or chiefly fulfils the requirement that science be a stable knowledge drawn from intelligible things. Indeed, theology is appropriately said to be 'God's science' in each of the four orders of causality. God is the formal cause of every knowable and knower, and God is the efficient cause of theology by the action of the Holy Spirit. Again, God is the subject-matter with which the science is concerned. Finally, and most importantly, God is the end towards which theology aims. It follows that theology is most truly wisdom because it is desired for its own sake without qualification.

The claim that the science of the divine is the goal of human knowing is reiterated under the fourth heading, but it receives its most important exposition in the questions of the treatise that follow the prologue. Here, Albert resolves a number of difficulties about the 'scientific' character of theology by explaining that it is a 'science according to piety' (*scientia secundum pietatem*).[50] The phrase is a variation on the Vulgate version of Titus 1:1 'according to . . . the apprehension of truth which is according to piety' (*secundum . . . agnitionem veritatis quae secundum pietatem est*). Albert invokes the phrase to describe a form of knowledge that instructs faith in the merit of certain deeds. This explicitly rhetorical knowledge aims to persuade by narrating particular events and by using as exemplars the actions of particular persons (*Summa Theologiae* 1.1.1 ad 1 (IAM 34: 6.62)). It does so because its audience contains many who can learn only by means of particulars, which function for them as 'universals in potency' (*potentia universalia*) (*Summa Theologiae* 1.1.1 ad 1 (IAM 34: 7.4-5)).

In saying this much, Albert has not by any means exhausted the phrase from Paul's Letter to Titus. Indeed, he assigns it central importance in the treatise by connecting it to two other doctrines. The first of these doctrines is Augustine's division in *De Doctrina Christiana* 1, of things into what signifies (*signum*), what is to be used (*uti*), what is to be enjoyed (*frui*), and what both uses and enjoys (*Summa theologiae* 1.1.3.1 (IAM 34: 9.70–10.20)). Albert extracts from this division the teaching that theology is a unified and separate science just so far as it attends to signs and other useful things as means to the enjoyable.[51] The second doctrine invoked by Albert asserts that the manner of theology, as exemplified in Scripture, ought to be multiply persuasive. Theology must then employ, not only the styles of affective rhetoric, but even the devices of the poets.[52] Thus Albert draws out of the Pauline phrase a description of theology as a form of knowledge of what can lead to salvation: that is, to the enjoyment of God.

The two doctrines are properly joined for Albert because he holds that genuine teaching about our highest end would have to try to move us towards it. He has in mind Aristotelian notions about the obligations of ethical discourse. So Albert notes Aristotle's concern for the limits on persuasion by mere argument for many hearers.[53] They must be taught, not by arguments, but by love or coercion. Indeed, because virtue is difficult for us, we need as many inducements to it as we can find, whether they are coercive or instructive. But Albert's tenet about the persuasive character of theology also derives from the view that all knowledge, speculative or practical, is ordered to human fulfilment in the vision of God.

The best known source for this view is Pseudo-Dionysius, who figures in the first treatise of the *Summa*. Albert relies particularly on the Dionysian assertion that the poetic devices of Scripture are so many 'coverings' (*integumenta*) or 'veils' (*velamina*).[54] But Albert knows very well that this assertion reaches beyond the text of Scripture to describe every creature and every science about creatures. Human sciences lead their learners by the hand through the progressively less material images until they are able to contemplate what is immaterial.[55] The language of Scripture recapitulates the pedagogy of the whole hierarchy of the sciences. The artful persuasions used in theology make explicit the teleology that underlies the Aristotelian trichotomy of the speculative sciences. The Aristotelian trichotomy turns out, once again, to be a somewhat misleading segment of a much larger and much richer hierarchy crowned by Scripture.

Albert makes this last point clear in a remarkable passage at the end of his discussion of theology. The passage is in the form of an objection, the body of which Albert both confirms and then underscores.[56] The objection narrates a sequence of motives that takes human learners through the arts and sciences. It supplies, as it were, the motive missing from Aristotle. The motives are all of them needs. Physical needs drive us to discover mechanical arts. The needs of the soul seek first for sciences of speech and logic as aids in further learning. Then they turn to physical bodies and their properties. But no knowledge of these can be had without knowledge of quantity of figure; so that the soul is driven to seek mathematics. Yet the principles of mathematics are not to be found within it. They can only be had in a higher science, which the philosophers call first philosophy or theology.

The objection wants to draw the conclusion that the highest science cannot have any end at all, or it would be for the sake of something further and so subservient. Albert replies that there are internal ends and external ends. The external end of theology is in the person who seeks to know its truth. The internal end is carried within the science itself. Thus theology is indeed, as the objection argued, the only free and untrammelled science. It is sought for its own sake. It is properly called 'wisdom' (*sapientia*), and it holds dominion over the other sciences. Their variety is required by the multiple weaknesses of human understanding, which needs a graded pedagogy in order to make progress towards apprehension of the simplicity of divine truth. Thus the hierarchy of sciences seems, from the vantage of theology, another instance of God's condescension. God reaches down by giving grace. The philosophers may be said to have a kind of 'revelation' through the intellect's light, but the actual contemplation of what is above and beyond the world of the senses requires the gracious giving of another light, which is here anticipated only by faith (*Summa theologiae* 1.1.4 ad 1 (IAM 34:15.41–59)). Without grace, human learning would be cruelly unfulfilled; without the divine light, human learning would be impossible.

Conclusions

The three texts discussed above serve to illustrate how Albert responded to characteristic tensions in the Aristotelian trichotomy of sciences. I have argued that he attempted to respond to these tensions by providing an additional narrative of the soul's ascent through the trichotomy and then explaining that narrative in terms of theology. In doing this, however, I have deliberately omitted one feature of Albert's responses. I will conclude by retrieving it and by suggesting that it points to a further range of tensions and difficulties.

In the last two texts discussed here, Albert includes in his description of our ascent up the hierarchy a requirement that our vocabulary become increasingly apophatic or 'negative' as one moves up the orders of learning to theological contemplation. So, in *De Intellectu*, Albert quotes philosophical authors in support of the assertion that the divine light 'has no name and cannot be told of' (*De Intellectu* 2.9 (Burgnet 9:517a)). In the *Summa* he lays out the entire Pseudo-Dionysian teaching about negative theology. This teaching is committed to the view that human language fails increasingly to capture the nature of 'reality' as one ascends the hierarchy. The most intelligible aspects of reality are hardest for us to understand and impossible for us to describe as they are in themselves. For this reason we must have recourse to apophatic language. This raises a profound question which does not invite an easy answer. Why, we might think, if our ascent up the orders of learning leads us to the realisation that ordinary language cannot capture the nature of the most intelligible reality, does such language provide us with accurate descriptions of the world, as that world is revealed to us in the different subjects of the hierarchy?

The simple way to capture these questions is to ask where Albert is standing when he speaks of the hierarchy. To this question it can be replied immediately that Albert is not really looking at the hierarchy. Rather, he is reading distinct, authoritative descriptions of the hierarchy. His first task as *magister* and as a teacher of theology is not to describe what he sees or experiences, but to join together what he and his students read. So the question raised by the requirement of negation for Albert's descriptions must be put more tellingly. Do the restrictions on language at the top of the hierarchy apply equally to descriptions of the top of the hierarchy as such? Again, is a description of the hierarchical position of metaphysics itself subject to the strictures of metaphysical language?

Albert does not explicitly address this or related formulations of the issue. Does his silence undermine the cogency of his account? A proper answer to this question would, I fear, take us into matters well beyond the remit of this paper and would require a proper treatment of his relation to Pseudo-Dionysius. That said, we might be able to save Albert from the appearance of a lack of rigour if we recall his use of language. I have noted at several points above that Albert's language becomes heavily charged whenever he describes the ascent to the top of the hierarchy. The pertinent passages of *De intellectu* are filled with lyrical quotations from numerous *auctoritates*: Pseudo-Dionysius, Augustine, Apuleius, to name but a few, and the corresponding parts of the other treatises repeat these remarks. In the *Summa*, the reader is given a doctrine about the rhetorical charter of theology. Because theology makes explicit the teleology of ascent, and because it offers the only sure way of ascending, theological language must be variously persuasive. So too are Albert's remarks on the hierarchy of the sciences. He does not offer them as explanations of theories but rather as part of a more general account of the mind's pursuit of God. Thus he tries from the opening of his paraphrase on the *Physics* to convert the Aristotelian trichotomy into an account of the mind's manners and motives in inquiring after God.

For Albert, then, the doctrine of the hierarchy of the sciences enables us to gauge properly the manners of knowing and the motives of learning. Through a description and classification of the individual sciences we unearth, he contends, the mutual connections and differences between different bodies of knowledge. This enables us to appreciate the nature of reality *both* as it appears in the world of the nature and in that world which exists beyond nature and the capacities of human comprehension. The study of natural sciences may lead the mind to the appreciation of a higher and more sublime form of study, which is ultimately to lead to the contemplation of the divine. It is here in a synthesis of science, philosophy, and theology – a synthesis which emphasises the autonomy of the physical sciences but which stresses their importance for the cultivation of a theological outlook – that the distinctiveness of high medieval science, and indeed of Albert's unique contribution to it, is to be seen.[57]

Notes

I gratefully acknowledge the support of a British Academy Post-Doctoral fellowship in Philosophy during which this paper was written. I am also happy to acknowledge the comments of Jill Kraye and Dag Nikolaus Hasse on earlier drafts of this paper. My greatest debt, however, is to Andreas Speer whose informed comments have helped me to appreciate in greater detail the subtleties of Albert's natural philosophy. This paper has also benefited from his excellent bibliographical advice.

1 In what follows I will render Latin term *scientia* as 'science'. I am very aware that a translation does the original term less than complete justice.
2 Two of the better recent books on medieval science which reflect these features are Lindberg 1992 and Grant 1996. Of further value is the section on 'Natural Philosophy' in Kretzmann, Kenny and Pinborg (eds) 1982.
3 For a general introduction to the intellectual history of this time, one which can help to orientate the reader towards a more specific understanding of phenomena such as 'science', 'philosophy' and 'theology' in the Middle Ages see Colish 1997.
4 Here one thinks of the characterisations of medieval philosophy peculiar to the first half of this century. See, for instance, Bertrand Russell's unworthy remark in Russell 1948: 484 ff., that medieval philosophers – his example being Aquinas – lacked the 'true philosophical spirit' because they sought to find arguments for conclusions given in advance.
5 For one of the best recent surveys of the different movements and schools of medieval philosophy see De Libera 1993. Other helpful histories are Marenbon 1991 and 1993; and Luscombe 1997. An older but still useful work is Copleston 1972.
6 On Albert's life and work see Weisheipl in Weisheipl (ed.) 1980a: 13–51, and Appendix 1: 565–77. Other important information on Albert's life and work can be found in Zimmermann and Vuillemin-Diem (eds.) 1981. Informative introductions to Albert's work can be found in De Libera 1990 and the articles by Sturlese 1997 and Anzulewicz 1999.

   Albert's works will be cited in one of two editions, to be noted as 'Borgnet' and 'IAM'. Borgnet refers to the *Opera Omnia* edited by Albert Borgnet in thirty-eight volumes (Albert: 1890–1899). IAM refers to the ongoing critical edition of the *Opera Omnia* edited by the members of the Institutitum Albertus Magni of Cologne (Albert: 1951–). The Borgnet edition will be cited by volume, page and column; the IAM edition by volume, page and line.
7 The most comprehensive history of thirteenth-century philosophy is Van Steenberghen 1991.
8 For an account of the psychological theories of the period see Dales 1995; and

for cosmology and astronomy see Grant 1994. Discussions of the principal areas of thirteenth century natural philosophy can be found in Grant 1981 and 1982; and Weisheipl 1985.

9 The methodological disputes within medieval mathematics are discussed by Clagett 1979.

10 For a discussion of the reception of Aristotle see Van Steenberghen 1955, Dod 1982 and Lohr 1982.

11 The main points of conflict between Aristotelianism and Christian theology concerned the following three substantive issues. The first focused on the creation of the world. A prominent feature of the Aristotelian cosmos was its eternity, defended by a variety of arguments in Aristotle's works. His position was that the cosmos did not come to be and cannot cease to be, since its elements have always behaved according to their natures. Consequently, there cannot have been a moment when the universe as we know it came into being, and no moment will come when it ceases to be; the universe is eternal. See *De Caelo*, Book 1: 10–11. From a Christian standpoint this is an intolerable conclusion. Not only does the account of creation in the opening chapters of the Book of Genesis claim that God created the heavens and the earth *ex nihilo*, but further, the idea of an active creator was fundamental to the Christian concept of God. Not surprisingly, among Aristotle's Christian commentators in the thirteenth-century, of which the most prominent are Thomas Aquinas, Bonaventure, Siger of Brabant (c. 1235–1282) and Boethius of Dacia (fl. 1260), we find a string of attempts to resolve this problem. For a full account of this debate see Dales 1990.

The second problem, also bearing on the relationship between Creator and creation, was that of determinism. The question of determinism in Aristotle's natural philosophy is a complex one. What needs to be stated here is that the universe as he described it contains unchangeable natures, which are the basis of a regular cause-and-effect sequence. Moreover, Aristotle's deity, the Prime Mover, is eternally unchanging and therefore incapable of intervening in the operation of the cosmos. The danger here is that within the Aristotelian framework no room could be found for miracles, especially as these occurrences are described in biblical literature. For a full discussion of determinism in Aristotle see Sorabji 1980. Such deterministic tendencies were viewed in some quarters as a challenge to Christian doctrine.

A third example of troublesome Aristotelianism concerned the soul. Aristotle had argued in his *De Anima* that the soul was the form of the body. It follows from this that the soul cannot have independent existence, since form, even if it can be distinguished from matter, cannot exist independently from matter. At death, therefore, when the individual dissolves, its form or soul ceases to be. Such a conclusion is clearly incompatible with Christian teaching on the immortality of the soul. The Aristotelian doctrine on the rational soul came down to the thirteenth century through the partial filter of the psychological theory developed by Averröes (1126–1198), as he had attempted to work out certain difficulties in Aristotle's epistemology. The full Averröistic theory, known as 'monopsychism' is complex. What is relevant here, is Averröes's claim that the immaterial and immortal part of the human soul, in the intellective soul, is not individual or personal but a unitary intellect shared by all humans. Thus immortality is preserved, but not personal immortality. Again the violation of Christian teaching is clear. For a discussion of these matters see Van Steenberghen 1980: 29–74; Walker Bynum 1995 Part III; and Dales 1995.

12 On Neoplatonic philosophy and Christian theology, see Marenbon 1993, Parts I–II.

13 The attempt successfully to appropriate ancient philosophy for the purposes of Christian theology goes back at least as far as the second century and to the two great theologians of the Alexandrian school, Clement of Alexandria (150–215 AD) and Origen (185–284 AD); for a discussion of their work see Stead (1994), especially Part II. The next influential band of theologians who had much of interest to say on the relations between philosophy and theology were the Cappadocian Fathers: Gregory of Nazianzus (330–89 AD), Basil of Caesarea (330–79 AD), and Gregory of Nyssa (335– c.95 AD). For a discussion of their work see Pelikan 1993. Of greatest importance to the medieval discussion of this issue, however, was Augustine, and in particular, his comments about ancient philosophy in *De Doctrina Christiana*. For a discussion of Augustine on this issue see Kretzmann 1990.

14 For a translation of the document bearing on the events in Paris at this time see Grant 1974: 42–4.

15 For the Latin text of this document see Chatelin and Denifle 1889–97, see vol. 1, 138 ff. For a discussion of the activities of the *magisteri* at Paris at this time see Glorieux (1933–4).

16 See Callus 1943 and Cobban 1988, see chapter 2.

17 It is important to note that so much of the new learning was transmitted to the universities by the new orders of friars, the Franciscans or 'Greyfriars', founded by St Francis of Assisi, and the Dominicans or 'Blackfriars', founded by St Dominic. For a helpful discussion of the place of the new mendicant with the worlds of arts, sciences, and letters of this time see Lawrence 1994.

18 On Grosseteste's scholarly career see McEvoy 1982 and Southern 1986. On Grosseteste's investigation of Aristotle's logic and its influence on subsequent scientific methodology see Crombie 1953 and Jeremiah Hackett, 'Roger Bacon on Scientia Experimentalis' in Hackett 1997. It is worth remarking in passing that Crombie's analysis of these issues is somewhat prone to overstatement.

19 Grosseteste's *Hexaëmeron*, which records some of the more salient aspects of his cosmology and cosmogony has recently been put into English in a fine translation by C. J. F. Martin, see Grosseteste 1996.

20 See Jeremiah Hackett's 'Roger Bacon: his life, career and works' in Hackett 1997.

21 Bacon's penchant for a 'proselytising' optics is recorded by Wiedermann 1914. A more sober evaluation of Bacon's work in optics is provided by Lindberg 1978 and 'Roger Bacon on light, vision, and the universal emanation of force' in Hackett 1997.

22 For further discussion of these aspects of Bacon's work see Easton 1952 and Crowley 1950.

23 A discussion of Bonaventure's general thinking on these issues can be found in Gilson 1924, see chapters I, XI, XII and XV. For less partisan discussion of Bonaventure's project in philosophy see Speer 1995 and 1997.

24 Even though Aquinas did not practice natural philosophy to the same extent as Albert, many of his more general thoughts in metaphysics do have implications for scientific methodology. For a discussion of these themes see Elders 1997 and Aertsen 1988, chapters 6–8.

25 Among the Neoplatonic texts we can include Albert's *Commentary on the Isagoge of Porphyry*, his *Commentary on the Liber de Causis (Book of Causes)*, and his *Commentary on the Divinis Nominibus (Divine Names)* of Pseudo-

Dionysis. For a discussion of Albert's debt to Neoplatonism see De Libera 1990, chapters 1–3.

26 As with Aquinas, it is vitally important to see Albert's philosophy, and likewise his interest in the hierarchy of the science, as arising in part from his responsibilities as a Dominican teacher of theology in one of that order's *studium generale*. It is also important to see Albert in the local context of the German Rheinland and, in particular, the philosophical movement, again heavily connected to the German Dominicans, that arose there in the mid thirteenth century. For discussions of these subjects see Sturlese 1981, and 1993: 324–6, and De Libera 1994.

27 See Weisheipl in Weisheipl 1980a.

28 For a list of these passages, and some reflections on their meaning see Weisheipl 1980b.

29 For a discussion of Aquinas as a commentator on Aristotle see Owens 1980 and Jenkins 1996.

30 On Albert's important psychological theories see: Gilson 1943; Michaud-Quentin 1955; Steneck 1980; Craemer-Ruegenberg 1981, and Sturlese 1993: 362–7. On Albert's debt to Avicenna see Vernier 1992 and Hasse 1997.

31 On Albert's use of multiple sources in the sciences see the essays by Dewan and Reeds in Weisheipl (ed.) 1980a.

32 Five books from Albert's treatise *De Animalibus* have recently been translated into English. They provide a good introduction to Albert's studies of animals and nature; see Albert 1987.

33 See Reeds in Weisheipl (ed.) 1980a.

34 An older but still useful discussion of Albert's use of experimental methods in science is Thorndike 1923–58, see vol. 2: 535 ff.

35 For an accessible account of these issues see Hankins 1982–9.

36 A helpful discussion of these issues can be found in Weisheipl 1985: 203–38.

37 See Weisheipl 1985: 203–38.

38 See for example, *Super Dionysium de Divinis Nominibus* 4 (IAM 37: 134.31–3).

39 *Physica* 1.1.1. (Borgnet 3:2a): '*Nihil ergo cadit in ratione mathematicorum de materia sensibili, sed potius de materia intelligibili quae est quantitas imaginabilis.*'

40 For further discussion of the 'mirroring' see Booth 1983: 165–80. For a more straightforward survey of Albert's explicit remarks on the nature of logic see Washell 1973.

41 For a discussion of Aristotle's scientific method and its possible deficiencies see Barnes 1975; and McKirahan 1992, see chapters 2–5.

42 It may also be noted that Albert wanted to find a place for the Pseudo-Aristotelian 'letter' that he mentions as his source of Book 1 (Borgnet 9). Weisheipl seems to think that the letter is the source for the whole, but the context suggests rather that Albert is using the letter as his guide on the question of emanation from the first cause. Cf. Weisheipl, 'Albert's works', in Weisheipl 1980a: 570, note 10d.

43 There are other more familiar formulations of the trichotomy as degrees of abstraction from matter in *De Intellectu*, as at 2. 6 (Borgnet 9: 512a–513b).

44 For a very helpful discussion of this aspect of Albert's work see Hasse 1999, especially 62 ff.

45 I follow the Latin version available in G. Camerarius (ed.) 1638, reprinted 1969) see 48–9, 49–53. 53–4, 56–62, and 63–4.

46 Avicenna, *Liber de Anima seu Sextus de Naturalibus*, S. Van Riet (ed.) 1968:

134–153. The *intellectus adeptus and the intellectus in effectu* are juxtaposed at 150.66–7.

47 On the authenticity and dating of the *Summa Theologiae*, see the editors' remarks in the *Opera Omnia*, IAM 34: v–xvi; and Tugwell 1988: 113, note 231.

48 The genre of *Summe Theologiae* peculiar to the thirteenth century is ably explained by Boyle 1982.

49 Psalm 138 Verse 6: 'Such knowledge is too wonderful for me, it is so high that I cannot attain it' (*Mirabilis facta est scientia tua ex me; confortata est, non potero ad eam*).

50 The phrase first occurs in *Summa Theologiae* 1.1.1 ad 1 (IAM 34: 6.61), and then recurs at 1.1.2 corpus (34: 8.47–8), 1.3.3 [objection 3] (34:13.46), and 1.1.5.1 ad 1 (34:16.37–8).

51 See, for example, *Summa Theologiae* 1.1.3.1 ad 3 (IAM 34: 11, 25–6). Cf. 1.1.3.2 ff.

52 Albert draws here on traditional lists of the *modi* or manners of speech in Scripture. He himself mentions nine *modi*: exemplary, perceptive, revelatory, hymnic, oratorical, parabolic, disputative, admonitory, and hortatory. See *Summa Theologiae* 1.1.5.4 [objection 13] (IAM 34: 21.25–41).

53 *Summa Theologiae* 1.1.5.4 ad 6 (IAM 34:21.76–79), with reference to *Nicomachean Ethics* 10.9 1179b4–21. Albert may be generalising Aristotle's point. Aristotle emphasises a distinction between the few and the many, and Albert too begins his reply by alluding to such a distinction. But it seems to disappear when Albert gets to the limits on persuasion. It seems for Albert that all human beings need to be treated as weak learners by the divine teaching.

54 *Summa Theologiae* 1.1.5.1 objection 1 (IAM 34:16.25), '*sub integumentis metaphoricis*', paraphrasing *Celestial Hierarchy* 2; 1.1.5.1 ad 2 (34:17.3–4), '*velaminum circumvalentum*', quoting *Celestial Hierarchy* 1.

55 *Summa Theologiae* 1.1.5.1 ad 2 (IAM 34:17. 22–31), quoting *Celestial Hierarchy* 1. Note especially the phrase '*materialis manuductio*.'

56 *Summa Theologiae* 1.1.6 objection 3 (IAM 34:23.22–49). The source cited is Aristotle, *Metaphysics* 1.2 982b29–30, but Aristotle provides no more than the statement that human nature is enslaved in many ways.

57 The effect of Albert's thinking upon later thinking in the Middle Ages is dealt with De Libera and Hoenen (eds.) 1995. Albert's influence upon the Renaissance is discussed by Mahoney 1980: 537–63.

Bibliography

Primary sources

Albert the Great (1890–99) *Opera Omnia*, ed. A. Borgnet, 38 vols, Paris: L. Vivès.

Albert the Great (1951–) *Opera Omnia*, ed. Members of the Institutium Albert Magni of Cologne, Munster i. W.: Aschendorff.

Albert the Great (1987) *Man and Beasts* (*De Animalibus Books 22–6*) trans. J. J. Scanlan, Binghampton N.Y.: Medieval and Renaissance Texts and Studies.

Alfarabi (1969) *Opera Omnia quae Latina Lingua Conscripta Reperir Potuerunt*, ed. G. Camerarius, Paris 1638, reprinted Frankfurt a. M.: M. Minerva.

Aristotle (1987) *The Complete Works*, ed. J.Barnes, 2 vols, Princeton: Princeton University Press.

Avicenna (1968) *Liber de Anima seu Sextus De Naturalibus*, ed. S. Van Riet, Louvain: Editions Orientalistes and Leiden: E. J.Brill.

Grosseteste, R.(1996) *On the Six Days of Creation*, trans. C .J. F. Martin, Auctores Britannici Medii Aevi, VI (2), Oxford: Oxford University Press.

Pseudo-Dionysius (1987) *The Complete Works*, trans. C. Luibheid (Classics of Western Spirituality), New York: Paulist Press.

Tugwell, S. (ed.) (1988) *Albert and Thomas: Selected Writings*, New York: Paulist Press.

Secondary literature

Aertsen, J. (1988) *Nature and Creature: Thomas Aquinas's Way of Thought*, Leiden: E. J. Brill.

Anzulewicz, H. (1999) 'Neuere Forschung zu Albertus Magnus: Bestandsaufnahme und Problemstellungen', *Recherches théologie et philosophie Médiévale* 65: 163–206.

Barnes, J. (1975) 'Aristotle's theory of demonstration' in J. Barnes, M. Schofield and R. Sorabji (eds) *Articles on Aristotle*, 4 vols., London: Duckworth, see vol. 1, *Science*, 68–87.

Booth, E. (1983) *Aristotelian Aporetic Ontology in Islamic and Christian Thinkers*, Cambridge: Cambridge University Press.

Boyle, L. E. (1982) *The Setting of the Summa Theologiae of Saint Thomas*, Etienne Gilson Series 5, Toronto: Pontifical Institute of Medieval Studies.

Callus, D. A. (1943) 'Introduction of Aristotelian learning to Oxford', *Proceedings of the British Academy* 29: 229–81.

Chatelin, A. and Denifle, H. (1889–97) *Chartularium Universitatis Parensis*, 4 vols., Paris: Delalain.

Clagett, M. (1979) *Studies in Medieval Physics and Mathematics*, London: Variorum.

Cobban, A. B. (1988) *The Medieval English Universities: Oxford and Cambridge to 1500*, Berkeley and Los Angeles: University of California Press.

Colish, M. (1997) *Medieval Foundations of the Western Intellectual Tradition 400–1400*, New Haven: Yale University Press.

Copleston, F. (1972) *A History of Medieval Philosophy*, New York: Harper and Row.

Craemer-Ruegenberg, I. (1981) I: 'Alberts Seelen – und Intelleklehre' in A. Zimmermann and G.Vuillemin-Diem (eds) *Albert der Grosse. Sein Zeit, sein Werk, sein Wirkung* (Miscellanea Medievalia 14), New York/Berlin: De Gruyter.

Crombie, A. C. (1953) *Robert Grosseteste and the Origins of Experimental Science, 1100–1700*, Oxford: Clarendon Press.

Crowley, T. (1950) *Roger Bacon: The Problem of the Soul in his Philosophical Commentaries*, Louvain: Nauwelaerts.

Dales, R. C. (1990) *Medieval Discussions of the Eternity of the World*, Leiden: E. J. Brill.

Dales, R. C. (1995) *The Problem of the Rational Soul in the Thirteenth Century*, Leiden: E. J. Brill.

De Libera, A. (1990) *Albert le Grand et la philosophie*, Paris: Vrin.

De Libera, A. (1993) *La philosophie médiévale*, Paris: Presses Universitaires de France.
De Libera, A. (1994) *La mystique rhénane. d'Albert le Grand à Maître Eckhart*, 2nd edn., Paris: Editions du Seuil.
De Libera, A. and Hoenen, M. (eds) (1995) *Albert Magnus und der Albertismus*, Leiden: E. J. Brill.
Dod, B. (1982) 'Aristoteles Latinus' in N. Kretzmann, A. Kenny, A. and J. Pinborg (eds) *The Cambridge History of Later Medieval Philosophy*, Cambridge: Cambridge University Press.
Easton, S. C. (1952) *Roger Bacon and his Search for a Universal Science*, Oxford: Blackwell.
Elders, L. (1997) *The Philosophy of Nature of St Thomas Aquinas*, Berlin and New York: Peter Lang.
Flasch, K. and Jeck, U. R. (eds) (1997) *Das Licht der Vernunft. Die Anfänge der Aufklärung im Mittelalter*. Munich: Beck.
Gilson, E.(1924) *La philosophie de Sainte Bonaventure*, Paris: Vrin.
Gilson, E. (1943) 'L'âme raissonable chez Albert le Grand', *Archives d'histoire doctrinale et litterature du Moyen Age*, 36: 5–72.
Glorieux, P. (1933–4) *Répertoire des maîtres en théologie de Paris au XIIIe Siecle*, 2 vols., Paris: Vrin.
Grant, E. (ed.) (1974) *A Source Book in Medieval Science*, Cambridge, Mass.: Harvard University Press.
Grant, E. (1981) *Studies in Medieval Science and Natural Philosophy*, London: Variorum.
Grant, E. (1982) *Much Ado About Nothing: Theories of Space and Vacuum from the Middle Ages to the Scientific Revolution*, Cambridge: Cambridge University Press.
Grant, E. (1994) *Planets, Stars and Orbs: The Medieval Cosmos 1200–1687*, Cambridge: Cambridge University Press.
Grant, E. (1996) *The Foundation of Modern Science in the Middle Ages: Their Religious, Institutional and Intellectual Contexts*, Cambridge: Cambridge University Press.
Hackett, J. (1997) *Roger Bacon and the Sciences: Commemorative Essays*, Leiden: E. J. Brill.
Hankins, J. (1982–9) 'Plato in the Middle Ages' in J. R. Strayer (ed.) *Dictionary of the Middle Ages*. 13 vols, New York: Scribners, see vol. 9: 694–704.
Hasse, D. N. (1997) *Avicenna's 'De Anima' in the Latin West, 1160–1300*, University of London Ph.D., unpublished.
Hasse, D. N. (1999) 'Das Lehrstück von den vier intellekten in der Scholastik: von den Arabischen quellen bis zu Albertus Magnus', *Recherches de Théologie et Philosophie Médiévale* 65: 21–78.
Jenkins, J. (1996) 'Expositions of the text: Aquinas's Aristotelian commentaries', *Medieval Philosophy and Theology* 5: 39–62.
Kretzmann, N., Kenny, A. and Pinborg, J. (eds.) (1982) *The Cambridge History of Later Medieval Philosophy*, Cambridge: Cambridge University Press.
Kreztmann, N. (1990) 'Faith seeks, understanding finds: Augustine's charter for Christian philosophy' in T. Flint (ed.) *Christian Philosophy*, Indiana: University of Notre Dame Press.

Lawrence, C. H. (1994) *The Friars: The Impact of the Mendicant Orders on Western Society*, London: Longman.

Lindberg, D. C. (1978) 'The science of optics' in D. C. Lindberg (ed.) *Science in the Middle Ages*, Chicago: University of Chicago Press.

Lindberg, D. C. (1992) *The Beginnings of Western Science: The European Scientific Tradition in Philosophical, Religious and Institutional Contexts, 600 BC to AD*, Chicago: University of Chicago Press.

Lohr, C. (1982) 'The Medieval interpretation of Aristotle' in N. Kretzmann, A. Kenny and J. Pinborg (eds) *The Cambridge History of Later Medieval Philosophy*, Cambridge: Cambridge University Press.

Luscombe, D. (1997) *Medieval Thought*, Oxford: Oxford University Press.

McEvoy, J. (1982) *The Philosophy of Robert Grosseteste*, Oxford: Oxford University Press.

McKirahan, R. (1992) *Principles and Proofs: Aristotle's Theory of Demonstrative Science*, Princeton: Princeton University Press.

Mahoney, E. P. (1980) 'Albert the Great and the *Studio Patavino* in the late fifteenth and early sixteenth centuries', in J. A. Weisheipl (ed.) *Albertus Magnus and the Sciences*, Toronto: Pontifical Institute of Medieval Studies.

Marenbon, J. (1991) *Early Medieval Philosophy*, London: Routledge.

Marenbon, J. (1993) *Later Medieval Philosophy*, London: Routledge.

Meyer, G. and Zimmermann, A. (eds) (1980) *Albertus Magnus. Doctor Universalis 1280/1980*, Mayence: Walberberger Studien 6.

Michaud-Quantin, P. (1955) 'Albert le Grand et les puissances de l'âme', *Revue de Moyen Age Latin* 11: 59–86.

Owens, J. (1980) 'Aquinas as Aristotelian commentator' in *St Thomas on the Existence of God*, Albany: University of New York Press.

Pelikan, J. (1993) *Christianity and Classical Culture: The Metamorphosis of Natural Theology in the Christian Encounter with Hellenism*, New Haven: Yale University Press.

Reeds, K. (1980) 'Albert on the natural philosophy of plant life' in J. A. Weisheipl (ed.) *Albertus Magnus and the Sciences*, Toronto: Pontifical Institute of Medieval Studies.

Russell, B. (1948) *A History of Western Philosophy*, London: Unwin.

Sorabji, R. (1980) *Necessity, Cause and Blame: Perspectives on Aristotle's Theory*. London: Duckworth.

Southern, R. (1986) *Robert Grosseteste: The Growth of an English Mind in Medieval Europe*, Oxford: Oxford University Press.

Speer, A. (1995a) 'Von der Wissenschaft zur Weisheit. Philosophie im Übergang bei Bonaventura' in T. Borche and J. Kreuzer (eds) *Weisheit und Wissenschaft*, Munich: J. P. Bachem Verlag.

Speer, A. (1995b) *Die entdecke Natur. Untersuchungen zu Begründungsversuchen einer 'scientio naturalis' im 12. Jahrhundert*, Leiden: E. J. Brill.

Speer, A. (1997) 'Bonaventure and the question of a medieval philosophy', *Medieval Philosophy and Theology* 6: 25–46.

Stead, C. (1994) *Philosophy in Christian Antiquity*, Cambridge: Cambridge University Press.

Steneck, N. H. (1974) 'Albert the Great on the classification and localization of the internal senses', *Isis* 65: 193–211.

Steneck, N. H. (1980) 'Albert on the psychology of sense perception' in J. A. Weisheipl (ed.) *Albertus Magnus and the Sciences*, Toronto: Pontifical Institute of Medieval Studies.

Sturlese, L. (1981) 'Albert der Große und die deutsche philosophische Kultur des Mittelalters', *Freibürger Zeitschrift für Philosophie und Theologie* 28: 133–47.

Sturlese, L. (1993) *Die deutsche Philosophie im Mittelalter. Von Bonifatius bis zu Albert der Große*, Munich: Beck.

Sturlese, L. (1997) 'Der Rationalismus Alberts des Grossen' in K. Flasch and U. R. Jeck (eds.) *Das Licht Der Vernunft. Die Anfänge der Aufklärung im Mittelalter*, Munich: Beck.

Thorndike, L. (1923–58) *A History of Magic and Experimental Science*, 8 vols, New York: Columbia University Press.

Van Steenberghen, F. (1955) *Aristotle in the West: The Origins of Latin Aristotelianism*, trans. L. Johnston, Louvain: Nauwelaerts.

Van Steenberghen, F.(1980) *Thomas Aquinas and Radical Aristotelianism*, Washington, D.C.: Catholic University of America Press.

Van Steenberghen, F. (1991) *La Philosophie au XIIIe Siecle*, 2nd edn, Louvain-Paris: Éditions Peeters, originally published in 1966.

Vernier, J. M. (1992) 'Le définition de l'âme chez Avicenne et s. Albert le Grand. Etude comparative du "Liber de anima" d'Avicenne at du "De anima" de s. Albert le Grand', *Revue des sciences philosophiques et théologiques anciennes et médiévales*, 76: 255–76.

Washell, R. F. (1973) 'Logic, language and Albert the Great', *Journal of the History of Ideas* 34: 445–50.

Walker Bynum, C. (1995) *The Resurrection of the Body in Western Christianity 200–1336*, New York: Columbia University Press.

Weisheipl, J. A. (ed.) (1980a) *Albertus Magnus and the Sciences*, Toronto: Pontifical Institute of Medieval Studies.

Weisheipl, J. A. (1980a) 'The life and works of St Albert the Great' in J. A. Weisheipl (ed.) *Albertus Magnus and the Sciences*, Toronto: Pontifical Institute of Medieval Studies.

Weisheipl, J. A. (1980a) 'Albert's works on natural science (*libri naturales*) in probable chronological order', Appendix I in J. A. Weisheipl (ed.) *Albertus Magnus and the Sciences*, Toronto: Pontifical Institute of Medieval Studies.

Weisheipl, J. A. (1980b) 'Albert's disclaimers in the Aristotelian Paraphrases', *Proceedings of the PMR Conference* 5: 1–27.

Weisheipl, J.A. (1985) *Nature and Motion in the Middle Ages*, Washington, D.C.: Catholic University Press of America.

Wiedermann, E. (1914) 'Roger Bacon und seine Verdienste um Die Optik' in A. G. Little (ed.) *Roger Bacon: Essays*, Oxford, Clarendon Press.

Zimmermann, A. and Vuillemin-Diem, G. (eds) *Albert der Große. Sein Zeit, Sein Werk, Seine Wirkung* (Miscellanea Medievalia 14), New York/Berlin: De Gruyter.

4 The seventeenth century and the reconstruction of knowledge

G. A. J. Rogers

Introduction

The seventeenth century offers probably the richest and most well trodden territory in both the history of science and the history of philosophy. While the line between the two areas was far from clearly drawn in the period, it is not too much of an exaggeration to say that both were characterised by intellectual revolutions which helped shape each for at least a century and perhaps for all time. Partly because of that, generalisations from the period are almost certain to encounter important counter-instances. If we can talk of directions at all the best that we can expect to identify will be trends, and even they would require us to look beyond the seventeenth century in both directions to be sure of their validity. Within that broader brief I shall focus for fairly obvious reasons on epistemic issues and the tensions which arose between the new sciences and the new epistemologies, partly induced by the widespread return to atomistic theories of matter, through the momentous developments in mathematical astronomy and mathematical physics, to central questions about method and the limits of knowledge. From at least the mid-sixteenth century the latter included the rising influence of sceptical argument, the clashes between these new forces and the authorised teachings as required by the churches and universities, and the powerful patronage which in general those institutions enjoyed.

My problem might be characterised as a question about the scope of natural philosophy. Did the method of natural philosophy, if there was such a method, apply to all possible forms of intellectual enquiry or was it confined to some sub-section? And similarly, was there some universal recipe for generating knowledge which could be applied to every kind of question? Or, conversely, is it the case that different areas required substantially different methods? Added to this there was the further question as to whether the same method generated the same degree of certainty, sufficient to count as *knowledge*, in all areas of enquiry. Perhaps some areas were inherently more problematic than others.

It is obvious enough that the most prominent area of potential conflict in the seventeenth century was that between the established institutions and their learning, on the one hand, encapsulated in the university syllabus but in much else besides, and, on the other, the achievements of the new science: the conflict between ancient and modern learning. The ancient learning, if we may call it such, more or less successfully blended with the accepted theology, or, since the religious upheavals of the sixteenth century, the accepted theologies, Catholic, Lutheran, Calvinist, with all their variations. It was therefore the potential conflict between the modern learning and the old theology that had in some way to be defused if serious damage was to be avoided to the collective intellectual edifice. And it was defusion rather than confrontation that was the chosen option of many in the first half of the century. Bacon, Galileo, Descartes are three prominent examples.

Defusion in their cases relied upon a redrawing of the map of knowledge in ways which avoided or played down the competitive nature of the knowledge claims made between natural philosophy and theology. But there were other options. One which was far from prudent was straightforward conflict. Bruno is a case in point. More problematic is Hobbes. A third possibility open to the natural philosophers was to carry the argument through from the natural sciences to theology; previously it had been more nearly the other way round. It was that third option which in England was more or less to carry the day and was never wholly absent from the scene. Its great exponents were Boyle, Locke and Newton, though they were only the most famous. It was this solution that was ultimately more stable than either of the other two and its triumph was an important feature, perhaps even central, to English society in the eighteenth century. In what follows I shall have little to say about the option of open conflict, mainly because it was in the short term at least very unsuccessful; the warfare of science and theology is a nineteenth century myth which need not delay us.

The new science encouraged a new epistemology. The pressures generated were of varying sorts and force. The implications of the revival of powerful atomistic theories of matter, their associated theories of light and perception, and their implications for epistemology is a story which still remains only partially told. With the abandonment of the scholastic perceptional story the most obvious casualty was the theory of sensible species, the transmitted forms without substance which guaranteed the veracity of perceptual judgements. The widespread adoption of some version of atomism (or corpuscularianism) was incompatible with such a story. But the atomism itself raised more questions than it answered about the objects of perception. It thus made more obvious the force of the sceptic's arguments, which in any case received independent support from the widespread perceived lack of agreement among the philosophers and theologians.

The trend in the period is roughly the following. Setting aside for the moment the problems raised by mathematical astronomy, in the earlier part of the century the New Science was not immediately seen to raise epistemic issues of a kind that threatened the programme. Bacon and Galileo, for example, did not regard the possibility of knowledge of the natural world as problematic. With the coming of the powerful influence of Cartesian subjectivism, however, there are important changes. The focus is now on the *individual's* search for truth, a shift so radical in its implications and ultimately so alien to the social dimensions of natural science that it could never be wholly adopted. At the same time we have the introduction of an account of matter and its properties which deprived, or threatened to deprive, sense experience of its assumed authority. The story, briefly, is this. In the standard accounts visual perception was explained by the transmission of the form of the object to the perceiver in the nature of visible species. In veridical perception, therefore, there was an exact match between the qualities of the object perceived and the qualities seen by the perceiver: how else could it be *veridical* perception at all? I shall call this the Resemblance Condition for veridical perception. With the introduction of corpuscular theories of matter this condition was to be abandoned. For the causes of perception – the colourless, odourless, invisible particles of classical atomism – were now taken to be quite different from their perceived effects, in the case of vision the coloured objects of ordinary perception. The abandonment of the Resemblance Condition by Galileo, Descartes and Hobbes was to generate strong tension between the search for certainty in natural philosophy and the empirical basis for that certainty. It was a tension compounded with the introduction by Descartes of the term 'idea' into his account of things. For the term seemed to authorise and consolidate an epistemic gap between the knower and the known that invited the serious attentions of the sceptic. The 'New Way of Ideas', as it was to be called by the end of the century, was to dominate philosophy for the best part of two centuries and always threatened the realist scientific programme which the philosophers of the seventeenth century believed they had established. It is the beginnings of that story in the seventeenth century, together with its interactions with religious belief, that will be my focus.

Bacon

I have already described Bacon as a defuser of conflict. He describes philosophy thus:

> The object of philosophy is three-fold – God, Nature and Man; as there are likewise three kinds of ray – direct, refracted, and reflected. For nature strikes the understanding with a ray direct; God, by reason of the unequal medium (viz. his creatures) with a ray refracted; man, as shown and exhibited to himself, with a ray reflected.
>
> (Robinson 1905: 453)

Bacon thereby distinguishes the direct empirical method for knowledge of nature from the 'refracted' knowledge of God. Both, however - indeed all three forms of knowledge - need not be in any way problematic. Bacon has no sceptical doubts about the possibility of empirical knowledge. Perceptual knowledge is 'natural' and direct. The post-Cartesian worries about the problematic nature of sense experience never features in Bacon's world. It is of course relevant to this that he never talks of ideas and his direct realism is closer in spirit to Aristotle than it was to Descartes or Locke.

Bacon then goes on to give us his version of the tree of knowledge:

> Philosophy many therefore be conveniently divided into three branches of knowledge: knowledge of God, knowledge of Nature and knowledge of Man, or Humanity. But since the divisions of knowledge are not like several lines that meet in one angle; but are rather like branches of a tree that meet in one stem (which stem grows for some distance entire and continuous, before it divides itself into arms and boughs); therefore it is necessary before we enter into the branches of the former divisions, to erect and constitute one universal science, to be as the mother of the rest.
>
> (Robinson 1905: 453–4)

Bacon explained the trunk, the *philosophia prima*, as the receptacle for all those axioms held in common by all the separate sciences.

An important feature of Bacon's tree of knowledge is that the various branches, while sharing the *philosophia prima*, are otherwise independent of each other. Knowledge of nature, therefore, does not presuppose knowledge of God. It thus allows for the autonomy of the natural sciences to this extent, that the truths of physics, say, can neither be inferred or conflict with the truths of theology. This was obviously very important for the natural philosophers. For it implies that they should not have constantly to be looking over their shoulder to see if their discoveries infringe the accepted theology.

A second feature of Bacon's tree should also be stressed. It is that Bacon obviously holds that theology can yield knowledge, at least in substantial areas. In fact, somewhat in conflict with the notion that theology and physics are different constituent branches of the tree of knowledge, he claims that natural theology is known in much the same way as physics. For divine philosophy or natural theology 'is that knowledge or rudiment of knowledge concerning God which may be obtained by the contemplation of his creatures' (*Advancement of Learning*, Robinson 1905: 91). The Book of Nature, therefore, was a source of knowledge of God's power and wisdom, and the study of that book would confirm, not undermine, religious conviction. Nor should we confine our investigations to the superficial properties of natural objects. Bacon clearly sees penetration into the inner workings of things as a source of religious confirmation. To this important extent natural philosophy as practised by Bacon's experimenters could be seen as a form of religious activity. Dissection as well as holy communion could be an act of worship.[1]

There was however an important difference between the divine and the mundane. For nature, though often reluctant to reveal her secrets, could be made to do so if suitably constrained. Just as the guilty prisoner could be forced to confess by means of torture, so could nature. The latter is called an experiment. In principle there were no limits to what can be known about nature. With theology, however, the matter was sometimes otherwise. There are religious mysteries beyond our comprehension. It is, furthermore, wrong to expect to discover philosophical truth in scripture, a mistake made by the Paracelsians. For the function of scripture is to inform us about matters of faith, manners, liturgy and government, not natural philosophy (cf. *Advancement of Learning*, ibid.: 174–5). If theology was not to reveal truths of natural philosophy, then it was equally mistaken to suppose natural philosophy could in general lead to theological knowledge: 'if any man shall think by view and inquiry into these sensible and material things, to attain to any light for the revealing of the nature and will of God' Bacon tells us in the *Valerius Terminus*, 'he shall dangerously abuse himself' (*Valerius Terminus*, ibid.: 186).

Bacon saw the pursuit of knowledge to be right and good, providing it met the condition that '*all knowledge is to be limited by religion, and to be referred to use and action*' (*Valerius Terminus*, ibid.: 186). It was not the pursuit of knowledge as such which had caused the angels and man to fall. In the case of the former there had been no intention to emulate God in his goodness but only in his glory. Similarly the fall of man was marked not by wishing to be like God in total but in respect of knowing good and evil. The picture which Bacon has about the proper scope of science, then, is keenly circumscribed by his religious convictions. With regard to the natural world, pursuing knowledge of it was part of our wider duty to benefit humanity. He tells us that he prays to 'God the Father, God the Son, and

God the Holy Ghost that remembering the sorrows of mankind and the pilgrimage of this our life wherein we wear out days few and evil, they will vouchsafe through my hands to endow the human family with new mercies' ('Author's Preface to *The Great Insaturation*', ibid.: 246). Such utilitarian objectives must not interfere with things divine and not lead to incredulity about the divine mysteries. They must 'give faith that which is faith's' and proceed, purged of the venom of the snake of Eden, to 'cultivate truth in charity' (ibid.: 247). Often Bacon appeals to the prelapsarian picture of the naked, innocent pursuit of truth as the model for his programme. It is this simple search for truth uncontaminated with our own fancies that should characterise the natural philosopher 'for God forbid that we should give out a dream of our own imagination for a pattern of the world' (ibid.: 253–4).

The twin dangers as Bacon saw it to his programme may well be understood as intellectual sins which flowed from a misidentification of the proper ambition of science. On the one hand there were those who claimed already to know the law of nature: 'For as they have been successful in inducing belief so they have been effective in quenching and stopping inquiry' and done much harm by stifling the efforts of others. But the sceptics, though they have important argument on their side, are equally at fault. Between the two extremes of 'the presumption of pronouncing on everything and the despair of comprehending anything' ('Author's Preface to the *Novum Organum*', ibid.: 256) there lies a middle way. Whether anything really can be known is to be settled not by argument but by *trying*. How that trying was to be accomplished is the whole Baconian programme. It was *activity*, not contemplation, that marked the path to nature.

Bacon wastes little time in his numerous writings in attempting to provide a moral or theological justification for his programme. To him and others it hardly seems necessary. And the arguments drawn from atomism and scepticism against the kind of direct realism that he more or less unquestioningly accepted had yet to receive formulations powerful enough to hold his attention. In a sense Bacon remains an Aristotelian in epistemology, partly reflected perhaps in his holding to the language of forms in his account of the objectives of scientific enquiry. Although he is generally regarded as leaning towards a corpuscularian theory, it was closer to Paracelsus than it was to Gassendi, and although he even talks of primary and secondary qualities, when he does so it is terms closer to Aristotle than to Locke.[2, 3] Nor, it is worth adding, did Bacon attempt to square his own semi-Paracelsian cosmology with the text of Genesis. As his latest editor writes: 'Bacon stripped Paracelsian materials from their Scriptural context, and did not try to legitimize cosmological doctrines by representing them as infallible readings of Genesis'.[4] With regard to scepticism, while he saw certain forms of scepticism as negative, Bacon was clear that the mitigated sceptics had an important point to make. For Bacon well knew that his own speculations yielded only provisional conclusions which may serve as 'wayside inns, in which the mind may rest and refresh itself on its journey to more certain conclusions' (*The Great Instauration. Plan of the Work*, Robinson 1905: 253). Although a full-blooded scepticism would be self-defeating, caution against claiming more than was justified was for Bacon an important virtue. But he never lost sight of the final goal of science as being a form of certainty, one which probably owed more to Aristotle than he was prepared to recognise, and which was quite different from that to which the mathematical astronomers and philosophers aspired. It is now time to turn to one of these.

Galileo

Bacon never found himself in serious conflict with either the church or the universities. That may in part be because he did not campaign strongly for any particular claims about the natural world and therefore in that respect he did not disagree with accepted positions. Sometimes he does not even appear to be advocating reform of the university curriculum, for he makes a point of saying that the standard scholastic offering has pedagogically much to commend it. The position was quite otherwise for Bacon's contemporary, Galileo. Although there is a sense in which Galileo became confrontational in his debate with the church, detectable, for example, in the opening paragraph of the *Dialogue Concerning the Two Chief World Systems* of 1632, at an earlier stage he was careful to underline that there was not, and could not be, conflict between the claims of theology and those of natural philosophy. The classic statement here was his 'Letter to the Grand Duchess Christina' of 1615, which followed a growing debate about the theological implications of the Copernican system in which Cardinal Bellarmine had come out against the Copernican supposition of a moving earth. The debate at this stage was primarily in terms of whether or not biblical texts could settle matters of astronomical truth. Galileo's approach was that, first, two truths could never be in conflict, so that if the demonstrations of the astronomers proved that the earth was indeed in motion then this showed that biblical texts which suggested otherwise had been misinterpreted. They should not be given the literal reading that the anti-Copernicans sought but, rather, should be read metaphorically, as many other passages in the Bible already were. Implicit in this was Galileo's commitment to the autonomy of the natural sciences. Conflicts between a literal reading of the biblical text and known physical facts can always be resolved once it is allowed that the purpose of the Bible is not to teach natural philosophy but to 'accommodate the minds of the very unrefined and undisciplined masses', from which 'one can very reasonably deduce that, whenever the same Holy Scripture has seen fit to assert any physical conclusion (especially on matters that are abstruse and difficult to understand), [the church] has followed the...rule [that wise interpreters formulate the true meaning] in order not to sow confusion into the minds of the common people' ('Letter to the Grand Duchess Christina' in Maurice A. Finocchiaro (ed.) 1989). Therefore, Galileo goes on, 'I think that in disputes about natural phenomena one must begin not with the authority of scriptural passages but with sensory experience and necessary demonstrations.' He claims that the work of the natural philosopher and the theologian stand on a par: 'For the Holy Scripture and nature derive equally from the Godhead, the former as the dictation of the Holy Spirit and the latter as the most obedient executor of God's orders' (ibid.: 93. Translation modified).

Galileo goes on to say that 'God reveals Himself to us no less excellently in the effects of nature than in the sacred words of Scripture' and it therefore follows that natural phenomena placed before our eyes or proved by necessary demonstration should not be called into question on account of scriptural passages which appear to have a different meaning. By implication, then, Galileo thus places the observations of the natural philosopher at least on the same level as those of the theologian and equally revealing of God's nature. It was a strong, even brave, claim to make in 1615. And it was also a commitment to the possibility of achieving truths in central areas of natural philosophy.

Galileo recognises that not all questions about the natural world admit of a certain answer. Some, such as whether the stars are animate, can only be a matter of probable opinion. Others, however, 'on the basis of experiments, long observations, and necessary demonstrations' can yield complete certainty. In the former case, where one has only opinion and faith, then 'it is appropriate piously to conform to the literal meaning of Scripture' (ibid.: 104). But when we can first attain the facts then we should do so. These truths of natural philosophy could even enable us better to obtain the true meaning of Scripture, and Galileo quotes St. Augustine in support.

Although mathematics is infinitely more important to Galileo than it was to Bacon, and although Galileo's atomism and commitment to the primary-secondary quality distinction is central to his account of the physical world and our knowledge of it, he shares with Bacon a belief that empirical investigation of the natural world can indeed lead to certainties about its nature. And for all his alleged Platonism, it is observation that lies at the basis of that certainty, albeit that the observation should so far as possible come in a quantified form. Further, he vigorously repudiates the position of Bellarmine that Copernicus was concerned only to save the phenomena: 'to claim that Copernicus did not consider the earth's motion to be true could be accepted perhaps only by those who have not read him'.[5]

For all that, the alleged Platonism does have some basis. Galileo's famous view that the book of nature is written in the language of mathematics, and that the characters of that language are the geometrical figures, implied that ultimate knowledge of the natural world would come in the form of Euclidean propositions, with all the certainty of that discipline. It is scarcely surprising that the works of Sextus Empiricus do not feature in Galileo's writings.[6]

Finally, with regard to Galileo, we may note that he shared with Bacon an optimistic belief that knowledge of the natural world could be indefinitely expanded. He also knew he had contributed towards that. As he put it: 'there has been opened up to this most excellent science, of which my work is merely the beginning, ways and means by which other minds more acute than mine will explore its remote corners' (Crew and de Salvio 1914: 153–4). It was a prophetic vision which saw no obvious boundaries to the enterprise of science.

Descartes

Like Bacon, Descartes, in the *Principles of Philosophy*, offers a tree of knowledge. But in important respects the trees differ. Descartes characterises his like this: 'The roots are metaphysics, the trunk is physics, and the branches emerging from the trunk are all the other sciences, which may be reduced to three principal ones, namely medicine, mechanics and morals.' By morals, Descartes tells us, he means 'the highest and most perfect moral system, which presupposes a complete knowledge of the sciences and is the ultimate level of wisdom' (Cottingham, Stoothhoff, Murdoch and Kenny 1985–91 (hereafter PW), vol. 1: 186). Metaphysics Descartes had already explained. It is the first part of philosophy 'which contains the principles of knowledge, including the explanation of the principal attributes of God, the non-material nature of our souls and all the clear and distinct notions which are in us.' It thus corresponds to the *Meditations* and Part I of the *Principles of Philosophy*. The second part of Descartes's tree, the trunk, is physics, which contains, first the principles of material things, and then 'the general composition of the entire universe and...the nature of this earth and all the bodies that are commonly found upon it, such as air, water, air, fire, magnetic ore and other minerals'. This part corresponds to the other three books of the *Principles*. The branches of the tree correspond to the all the other subjects: the individual nature of plants, animals, and, above all, human beings. We need to know the nature of the last especially so that we can both discover and make best utilitarian use of the remaining sciences. Descartes, like Bacon, though not so loudly, is committed to knowledge bringing with it human benefits, an objective he was often to underline.

The criterion of clear and distinct ideas, so central to Descartes's account of knowledge, implies that where we have no such ideas then we can have no knowledge. Obtaining such ideas depends on beginning from the most clear and simple possible. The analysis of the compounds into the epistemically simple is the first step towards knowledge. With one important exception Descartes seems committed to saying that in principle all that we can think about or encounter can be so analysed. The exception is our idea of God. There is a very good reason why we can never have anything more than a partial, and therefore limited, idea of God and that is because God, uniquely, includes a positive idea of infinity. Part of what is meant by the idea of God involving a positive idea of infinity is that in the case of God alone we see that there is no attribute which He lacks or which He has merely potentially. Since He is perfect there is necessarily no attribute of a positive kind which could be added to Him. This distinguishes God from all other infinities which are merely negative. But an infinite positive idea is, as a matter of logic, something that a finite mind cannot grasp. To that extent, at least, therefore, theology can at best be only a partial human science. Thus, although my idea of God may be 'the truest and most clear and distinct of all my ideas' it is never other than 'inadequate' (Third Meditation, PW2: 32).

Closely linked to Descartes's positive idea of God's infinity is his belief that it is impossible for human beings to infer from God's creation to God's intentions. Because God's nature is 'immense, incomprehensible and infinite' and 'capable of countless things whose causes are beyond my knowledge', he writes, 'I consider the customary search for final causes to be totally useless in physics' (Fourth Meditation, PW2: 39). That God's intentions are inscrutable to the natural philosopher – we will recall that physics covers all the objects on this earth and beyond – could be seen as a rejection of natural theology as a legitimate intellectual enquiry. It was so seen by Gassendi who responded strongly: 'there is an obvious danger that you may be abandoning the principal argument for establishing by the natural light the wisdom, providence and power of God, and indeed his existence', Gassendi wrote, and there is much more in the same vein (Fifth Set of Objections, PW2: 215). Descartes responded: God's created objects, the various parts of plants and animals, for example, may well lead us to admire God as their efficient cause, but they give no clue as to God's purpose in making them. Descartes did not just resist the standard natural theology, he was very keen to do so. For it was crucial to his objective of removing final causes from natural philosophy, a vital step in the vanquishing of scholastic philosophy.

It is well known that Descartes's physics makes use of a priori argument to move from the immutability of God to his first laws of nature: the principle of straight-line inertial motion. So there is this strong rationalist element in his physics. But it is also clear that he recognised that much about the natural world could only be discovered by observation and experiment. When we reach the limits of such empirical discovery then we are thrown back on hypothetical explanation: 'With regard to the things which cannot be perceived by the senses', he writes in the closing sections of the *Principles*, 'it is enough to explain their *possible* [my emphasis] nature, even though their actual nature may be different' (*Principles* IV. 204, PW1: 289). But this hypothetical model, in so far as it produced an explanation for all the known phenomena of nature, thereby acquired more than a moral certainty. For how could so much be accounted for if the explanation was insecure? The Principles may even be allowed as absolute certainties when we recall that 'they have been deduced in an unbroken chain from the first and simplest principles of human knowledge' (*Principles* IV. 206, PW1: 290). Descartes was at least as optimistic as Bacon and Galileo about the possibility of a universal science of nature. He was also keen to avoid theological dispute arising from his philosophy (though unsuccessfully), as his prefatory Letter to the Doctors of the Sorbonne in the *Meditations* testifies.[7]

I have already suggested that Descartes's introduction of talk of ideas into his epistemology was important, no doubt inadvertently.[8] It is a paradox of seventeenth-century thought that the thinker who was most keen to defeat uncertainty should have left such an inheritance. I suggested that Descartes's abandoning of what I call the Resemblance Condition as a necessary condition for veridical perception was central to that. He explains it in the *Optics* by analogy with a blind man feeling his way by means of a stick. Nothing is transmitted from the ground to the hand of the man but he nevertheless comes to know its properties: 'hence you will have reason to conclude that there is no need to suppose that something material passes from objects to our eyes to make us see colours and light, or even that there is something in the objects which resembles the ideas or sensations he has of them' (*Philosophical Works* vol. 1: 153). Such an account has the great advantage, Descartes points out, of removing the need for 'intentional species'. But the problem which Descartes thereby left to his successors was a very real one: if there is no resemblance between ideas and their causes, how may we know what the natures of those causes actually are, or, indeed, if there are any such external causes at all? It was of course an issue to remain henceforth at the centre of philosophy.

Hobbes

Bacon and Descartes aspired to reform learning and provide a universal method for obtaining knowledge. At least as original and certainly much more confrontational, so did Hobbes. But for all his apparent willingness to test the tolerance of church, state and universities Hobbes was very careful in his account of philosophy to indicate that there were certain areas which lay outside its scope. Thus, although he laid claim to being the first to make politics a science and identified himself with Copernicus, Galileo and Harvey, Hobbes was careful to argue in his account of philosophy that it followed from his epistemological principles that there were real limits to philosophy which importantly excluded most interesting questions of theology.[9] And by philosophy Hobbes understood knowledge generated by a particular method, a method which had its origins in experience and rational inference. He explained it all several times (more or less consistently) but perhaps nowhere more clearly than in *De Corpore*.

Philosophy Hobbes defined as: 'such knowledge of effects or appearances, as we acquire by true ratiocination from the knowledge we have first of their causes or generation: And again of such causes or generations as may be from knowing first their effects' (Molesworth 1839: I: 3). The difference between philosophy and ordinary empirical knowledge he explained like this: 'although sense and memory of things, which are common to man and all living creatures, be knowledge, yet because they are given to us immediately by nature, and not gotten by ratiocination, they are not philosophy'. The two ingredients of philosophy, therefore are sense experience (the objects of which Hobbes calls 'phantasms') plus deduction (which Hobbes calls 'ratiocination').[10] Philosophy requires both experience and reason. It follows from this that where there is no experience there can be no philosophy. And it was precisely this that limited its scope. The subject matter of philosophy, Hobbes tells us, 'is every body of which we can conceive any generation and which we may . . . compare with other bodies' in virtue of their known properties and generation. It therefore excludes theology because God is eternal and ungenerable, incomprehensible, indivisible and uncompoundable. It also excludes 'the doctrine of angels and all such things as are thought to be neither bodies nor properties of bodies' (Molesworth 1839: I: 10). His list of exclusions continues. History 'as well *natural* as *political*' is out because it is based on experience only, not ratiocination. And so is all knowledge acquired by divine inspiration; all false doctrines, such as astrology (this is Hobbes writing in 1649); and finally God's worship, which is transmitted only by the authority of the Church and is the object of faith, not knowledge.[11] Later in his chapter 'Of the World and Stars' he makes explicit that all talk of infinities is excluded from knowledge and therefore from philosophy: 'the knowledge of what is infinite can never be known by a finite enquirer'. But whether or not the world is infinite makes no difference to us, Hobbes says, because whether the Creator has made it one or the other has no effect on what we perceive, for 'the same things which now appear, might appear, whether the Creator had pleased it should be finite or infinite' (*De Corpore*, EW I: 412). Such questions are to be decided by 'those who are lawfully authorised to order the worship of God'.

It is tempting to read Hobbes as if he were some latter-day positivist or even a Humean, committed to a verificationist theory of meaning. Or, alternatively, since he believes that all knowledge includes ratiocination, that he is some kind of deductivist with all scientific knowledge flowing from the first definition of terms. On the first charge, we must say that there is no textual evidence of which I am aware that forces us to the strong conclusion about Hobbes' positivism in natural philosophy. It would in any case be a difficult conclusion to draw, simply because he never made the mistake (prudential mistake) of ever arguing that the claims of the theologians were meaningless because unverifiable. Rather, the claims of the theologians were often undecidable by either reason or observation. As he explained it with reference to the words *infinite* and *eternal* 'of which we have in our mind no idea' (EW I: 414), we always speak something absurd and should therefore remain silent. And he adds, significantly, 'Whatsoever therefore is true, young geometricians think demonstrable; but elder not. Wherefore I purposely pass over the questions of infinite and eternal', contenting himself with acceptance of the word of Scripture and the custom and laws of his country. He was, in that sense, much closer to Popper than to Ayer.

Earlier I quoted Hobbes as saying that we have knowledge of things by sense 'given to us immediately by nature' (EW I: 3). Although Hobbes obviously subscribed to a causal theory of perception in which motion is transmitted from object to brain, and he also drew the same distinction as Descartes with respect to the Resemblance Condition (a matter about which he claimed priority and which led to an important quarrel with the Frenchman) Hobbes never introduced the Cartesian conception of ideas into his philosophy (though he occasionally used the term), which was anyway excluded by his materialism.[12] In that sense Hobbes's philosophy did not raise the same sceptical problem that Descartes's system did. Or, to put it another way, Hobbes's philosophy solved the Cartesian problem by not allowing its formulation. But Hobbes's solution would only be acceptable to materialists, who, in the seventeenth century, were, overtly at least, thin on the ground.

Hobbes was well aware that much of his philosophy was enormously controversial. For that reason his positive influence was rarely acknowledged by his contemporaries. For all that, I suspect it was in fact very considerable.[13] For example, *De Corpore* seems to have been a significant influence on the young Newton.[14] In a way Hobbes greatly expanded the scope of science, rather than otherwise, because it is he, perhaps above all, who showed in *Leviathan* how it might be extended into new territories. Finally on Hobbes, it should be noted that the activity of natural philosophy – and for Hobbes it is very much an activity – is justified by its utilitarian benefits in a way which befits the one-time amanuensis of Bacon. 'The end of knowledge is power' for 'the performance of some action' and the end of philosophy is that 'we may make use to our benefit of effects formerly seen . . . for the commodity of human life' (*De Corpore* EW I: 7).

Boyle

Robert Boyle shared with Hobbes a surprisingly large number of beliefs and assumptions about the natural world and our possible knowledge of it. But, while Hobbes managed to raise the hackles of almost every Christian in the land, Boyle had quite the opposite effect. Of his religious commitment there can be no question. Equally surely he saw himself as working within a programme created by Bacon. But he did not always follow Bacon in methodological matters and he differed from him in the kind of certainty that he expected the new natural philosophy to achieve. Bacon had aspired to certainty, though he never meant by that the kind of metaphysical or logical certainty which motivated the great rationalist thinkers later in the century. But Bacon's hope had become subtly modified by his successors into a more qualified claim to knowledge.[15] In the case of Boyle he was quite prepared to settle for something considerably less than certain knowledge in natural philosophy. It was 'The Excellency and Grounds of the Mechanical *Hypothesis*' that he attempted to establish, not its certain, necessary truth. His account of the properties of objects was based on a 'corpuscular hypothesis', not a certain truth; his debate with Hobbes about the implications of his vacuum pump experiments was based on what he took to be a plausible hypothesis about the causes of the spring and weight of the air, based on experimentally established facts, not an absolute certainty derived from necessarily true premises. His dispute with Spinoza about the certainty to be expected from natural philosophy rested on the presumption that absolute certainty was not a reasonable aspiration in most areas of natural philosophy.[16] Boyle's attack on the Paracelsians' account of chemical properties in the *Sceptical Chymist* was in part aimed at their too easy assumption of the truth of their premises without sufficient empirical evidence.

Although it would be wrong to see Boyle as either the only or even the first English natural philosopher to move away from an over-optimistic view about the certainty that natural philosophy could deliver in its results (Walter Charleton and Henry Power and, later, Joseph Glanvill, would be other instances, but there were many others), the whole tone of his writings and his actual laboratory method ran quite contrary to an expectation of quick and certain conclusions. In this he was closer to the spirit of Bacon's programme than Bacon's own words always suggested. For we will recall that Bacon saw the implementation of his instauration as requiring cooperative effort and moving from the tentative conclusions of the 'first vintage' in the 'wayside inn' to final certainties only over some period of time.

Boyle was quite clear that the study of nature was or could be an act of piety and it is well enough known that he was enormously keen to demonstrate the strong connection between natural philosophy and religion. Not only did the study of nature lead to a firmer knowledge of God's power and benevolence but it also led into a deeper moral knowledge of our purposes here on earth, an outlook he shared with virtually all the great figures of science of the late seventeenth century. But there were very real limits to our powers of comprehension. Our condition is such, Boyle tells us, that 'God intended the mind of man [to be] but of a limited capacity'. It is fitted 'to the attainment, though not of the perfect knowledge of truths of the highest orders, yet to the competent knowledge of as much truth as God thought fit to allow our minds in their present (and perchance lapsed) condition, or state of union with their mortal bodies' (Stewart 1979: 239).

Locke

The most important philosophical statement of the middle way to match our mediocre state was Locke's *Essay Concerning Human Understanding*. His view of the scope of natural philosophy was clearly indicated in his programme of education. From his analysis of the nature of the human intellect Locke was quite sure that the investigations of the natural philosophers are unlikely to bring us certain knowledge. Of natural philosophy, Locke writes:

> Though the World be full of Systems of it, yet I cannot say, I know any one which can be taught a Young Man as a Science, wherein he may be sure to find Truth and Certainty, which is what all Sciences

> give an expectation of. . . . I think the systems of *Natural Philosophy*, that have obtained in this part of the World, are to be read, more to know the *Hypotheses*, and to understand the Terms and Ways of Talking of the several Sects, than with hopes to gain thereby a comprehensive, scientifical, and satisfactory Knowledge of the Works of Nature.[17]
>
> (Yolton and Yolton 1989: 247)

Time and again in the *Essay* Locke emphasises that we are provided with just those faculties that will, if properly used, deliver all the knowledge that we require for this life, but we could never hope to obtain a comprehensive science of nature. We must of course remember that for Locke science is *scientia* and knowledge, with one important exception, is limited to that which is intuitively certain or derived by deduction from that which is intuitively certain. And, as with Hobbes, the certainty relates to what we have experienced, the ideas we have actually had. The knowledge we can have includes moral knowledge as well as the more obvious cases such as mathematics, but it stops short of much of natural philosophy. He expressed himself in a way which shows that Locke accepted a position which was, I think, very close to that of Boyle or Glanvill: 'the Systems of *Physicks*, that I have met with, afford little encouragement to look for Certainty or Science in any Treatise, which shall pretend to give us a body of *Natural Philosophy* from the first Principles of Bodies in general...', he writes, suggesting that it is *completeness*, in the sense of having been demonstrated from first principles, that he would see as a necessary condition for such a science (*Education*: 248). And Locke comes to this conclusion even though he allows that Newton has achieved something like this in some 'particular Provinces of the Incomprehensible Universe'. A central aspect of Locke's understanding of the possibility of achieving that comprehensive system was a consequence of his empiricism. The universe was composed of particles of matter too small to be detected by even the most powerful microscope. 'I am apt to doubt that, how far soever humane Industry may advance useful and *experimental* Philosophy *in physical Things*, *scientifical* will still be out of our reach: because we want perfect and adequate ideas of those very Bodies, which are nearest to us, and most under our Command' (Nidditch 1975, IV. III. 26: 556–7). Our ideas are acceptable for 'common Use and Discourse' but cannot generate 'scientifical Knowledge'. In such enquiries we are constantly forced back to analogical inference; probabilities are all we can hope for.

Nor did Locke think this was merely a passing phase. It was inherent in our condition. And it is important, I believe, for a proper understanding of Locke to appreciate that he operates with a hierarchy of knowledge and probabilities which entirely matches his picture of our place in the scheme of things. We have intuitive knowledge of our own existence, demonstrative knowledge of God, and the limited possibility of intuitive and demonstrative knowledge in ethics. Granted that our reason for being on this earth is to live a good life in order that we may deserve to share eternity with our Maker, knowledge of nature, though desirable, is not a *sine qua non* of fulfilling our terrestrial purpose. The knowledge we need for that is already to hand. Limited knowledge of nature is merely a bonus.

Newton

Despite the real influence of More and Cudworth in his Cambridge days Newton, like Locke – perhaps, as I have argued, *following* Locke – expressed his natural philosophy and the limitations of natural philosophy from within an empiricist epistemology.[18] Not only in his published works, the later editions of the *Principia* and the *Opticks*, but also in his manuscript drafts for many of his writings, and his private notebooks, Newton showed both a belief that what knowledge we have is dependant on experience and also a recognition that the only route to knowledge of general truths about the natural world was through induction, with its concomitant hazards, a state of affairs which he was quite happy to acknowledge. But Newton took this one stage further than most of his colleagues. For he was quite sure that having once identified the best possible method for reaching knowledge of matters of fact, as no doubt he believed that he, with Locke, had achieved, he turned that method towards the great religious questions that had come more and more to hold his attention. I will give one example of this but I think it sufficiently clear to make my point.

Amongst the books of the Bible that Newton especially wished to understand were the apocalyptic texts and he set down some methodological rules for interpreting them that bear more than a passing resemblance to the *Regulae Philosophandi* of the *Principia*. In expounding them he makes direct comparison with understanding the natural world. One of these methodological rules is an appeal to the principle of simplicity, which corresponds to Rule II of the *Principia* and reads as follows:

> Choose those constructions which without straining reduce things to the greatest simplicity. . . . Truth is ever to be found in simplicity, and not in the multiplicity, and confusion of things. As the world, which to the naked eye exhibits the greatest variety of objects, appears very simple in its internall constitution when surveyed with a philosophic understanding, and so much the simpler by how the better it is under-

> stood, so it is in these visions. It is the perfection of God's works that they are all done with the greatest simplicity. He is the God of order and not of confusion. And therefore as they that would understand the frame of the world must indeavour to reduce their knowledg to all possible simplicity, so it must be in seeking to understand these visions.
>
> (Yahuda MS 1 in Manuel 1974, Appendix A: 120)

So Newton is using the method of natural philosophy to understand the sacred texts. We have come full circle from those who sought to understand the true structure of the universe from the book of Genesis. It is a good measure of the change in the scope of science that was witnessed by the seventeenth century. And we have also moved from the goal of *scientia* in natural philosophy. Aristotelian and Baconian certainties, followed by the mathematically orientated ones of Galileo and Descartes, have given way to the hypotheses of Boyle, Locke's probabilities and analogies, and Newton's inductions. It is scarcely surprising that eighteenth century philosophy took the form that it did.

Notes

1 Bacon seems to believe, at least sometimes, that it was religious *wonder*, not *knowledge*, that flowed from natural philosophy.

2 Bacon writes in the *Novum Organum* of the 'primary forms' in nature and 'secondary qualities of matter'. He lists these as: 'attraction, repulsion, attenuation, conspissation, dilatation, astriction, dissipation, maturation, and the like'. Robinson 1905: 273.

3 Bacon in various places has quite a lot of interesting things to say about atomism, but he is careful not to side wholeheartedly with it. Thus he commends Democritus for not assuming that the properties of individual atoms are like the properties of gross bodies; they are 'quite different from anything subject to the senses' whose nature is 'entirely dark and secret'. (Rees 1996, *De Principiis atque Originibus*: vol. 6: 201). Bacon goes on to say that the basic properties of atoms must be quite unlike those of gross bodies, for the properties of the latter are the product of *combination*.

4 Rees (1996) xlviii. The quotation continues: 'The Paracelsians were wrong when they pretended to find all philosophical truth in the Scriptures and blackguarded other philosophies as heathenish and profane. There was, Bacon believed, no such enmity between God's word and his works.'

5 In a letter to Monsignor Dini in 1616 (Finocchiaro 1989: 60).

6 Even if the mathematical forms have a special place in Galileo's account of reality there is no evidence of a wider commitment to the moral and political aspects of Plato's philosophy or to its full-blown metaphysics. Galileo was not Plato or even Ficino, even if he was in important ways in their line of descent.

7 Despite the radical challenge that his thought presented to traditional learning there is no evidence of which I am aware that Descartes remained anything other than a practising communicant of the Catholic church.

8 The terminology was not new with Descartes but I suspect the use of it was. On this issue see Ariew and Grene 1995: 87–106.
9 On Hobbes' argument, see Molesworth 1839, vol.1: viii.
10 Hobbes calls it ratiocination which he identifies with 'computation', itself nothing other than 'addition and subtraction'. Logic for Hobbes includes not only simple inferences involving two terms but also syllogistic inference involving three. Cf. *Elements of Philosophy*, part I, EW I: 1–90.
11 Part I of *De Corpore* was completed in Paris in 1649, confirmed by Hobbes.s letter to Samuel Sorbière from Paris on 14 June of that year. Cf. Malcolm 1994, vol. 1: 176.
12 For some of the implications of the quarrel between Hobbes and Descartes see Richard Tuck (1988), 'Hobbes and Descartes' in Rogers and Ryan 1988. Tuck's paper should also be read in conjunction with Schuhmann 1995.
13 Cf. G. A. J. Rogers (1988) 'Hobbes's hidden influence', in Rogers and Ryan 1988: 189–205.
14 On this see McGuire and Tamny 1983, especially 25, 219–21.
15 On this whole topic see Leeuwen 1970 and Shapiro 1983.
16 Cf. Hall and Hall 1964: 241–56.
17 Despite important changes Locke had expressed similar reservations in the original letter to Edward Clarke on which *Education* (1693) was based. Cf. de Beer 1976, vol.2: 785.
18 Cf. Rogers 1978 and 1979 in Yolton 1990: 339–81, and Rogers 1998 chapters VII and XII.

Bibliography

Ariew, R. and Grene, M. (1995) 'Ideas in and before Descartes', *Journal of the History of Ideas*: 87–106.

Cottingham, J, Stoothhoff, R., Murdoch, D. and Kenny, A. (eds) (1985–91) *The Philosophical Writing of Descartes*, 3 vols, Cambridge: Cambridge University Press.

Crew, H. and de Salvio, A. (trans.) (1914) *Two New Sciences*, New York: Macmillan.

de Beer, E. S. (ed.) (1976) *Locke: Correspondence*, Oxford: Clarendon Press.

Finocchiaro, M. A. (ed.) (1989) *The Galileo Affair: A Documentary History*, Berkeley, Los Angeles and London: University of California Press.

Hall, A. R. and Hall, M. B. (1964) 'Philosophy and natural philosophy: Boyle and Spinoza', *L'aventure de l-esprit. Melange Alexandre Koyré II*, Paris: Hermann.

Leeuwen, H. G. van (1970) *The Problem of Certainty in English Thought 1630–1690*, The Hague: Martin Nijhoff.

McGuire, J. E. and Tamny, M. (1983) *Certain Philosophical Questions. Newton's Trinity Notebook*, Cambridge: Cambridge University Press.

Malcolm, N. (ed.) (1994) *The Correspondence of Thomas Hobbes*, Oxford: Clarendon Press.

Manuel, F. E. (1974) *The Religion of Isaac Newton*, Oxford: Clarendon Press.

Molesworth, W. (ed.) (1839) *The English Works of Thomas Hobbes*, reprinted London and Bristol 1992: Routledge and Thoemmes.

Nidditch, P. H. (ed.) (1975) *Essay Concerning Human Understanding*, Oxford: Clarendon Press.

Rees, G. (ed.) (1996) *The Oxford Francis Bacon*, vol. VI (*Philosophical Studies c.* 1611– *c.* 1619), Oxford: Clarendon Press.

Robinson, J. M. (1905) *The Philosophical Works of Francis Bacon* (ed. from the edition of Spedding and Ellis), London: Routledge.

Rogers, G. A. J. (1978) 'Locke's *Essay* and Newton's *Principia*', *Journal of the History of Ideas*, now reprinted in J. W. Yolton (ed.) (1990) *Philosophy, Religion and Science in the 17th and 18th Centuries*, Rochester and Woodbridge: University of Rochester Press.

Rogers, G. A. J. (1979) 'Locke, Newton and the Cambridge Platonists on innate ideas', *Journal of the History of Ideas*, now reprinted in J. W. Yolton (ed.) (1990) *Philosophy, Religion and Science in the 17th and 18th Centuries*, Rochester and Woodbridge: University of Rochester Press.

Rogers, G. A. J. (1998) *Locke's Enlightenment*, Olms: Hildesheim.

Rogers, G. A. J. and Ryan, A. (eds) (1988) *Perspectives on Thomas Hobbes*, Oxford: Clarendon Press.

Schuhmann, K.(1995) '*Le Short Tract*, première œuvre philosophique de Hobbes', *Hobbes Studies*, vol VIII.

Shapiro, B. J. (1983) *Probability and Certainty in Seventeenth-Century England*, Princeton: Princeton University Press.

Stewart, M. A. (ed.) (1979) *Selected Philosophical Papers of Robert Boyle*, Manchester: Manchester University Press.

Yolton, J. W. and Yolton, J. S. (eds) (1989) *Some Thoughts Concerning Education*, Oxford: Clarendon Press.

Yolton, J. W. (ed.) (1990) *Philosophy, Religion and Science in the 17th and 18th Centuries*, Rochester and Woodbridge: Rochester University Press.

5 The end of hierarchy

Physics and metaphysics in the scientific revolution

J. R. Milton

I

At least among philosophers, one of the best-known arguments for the existence of God is the one set out by Descartes in the third Meditation. The argument is a causal one, starting from Descartes's own idea of God and proceeding to enquire into what the cause of this idea must be. Quite clearly some strong metaphysical principles are going to be needed if any such train of argument is to reach the desired conclusion, and Descartes showed no obvious reluctance in revealing at least one of them:

> It is manifest by the natural light that there must be at least as much reality in the efficient and total cause as in the effect of that cause. For where, I ask, could the effect get its reality from, if not from the cause? And how could the cause give it to the effect unless it possessed it? It follows from this both that something cannot arise from nothing, and also that what is more perfect – that is, contains in itself more reality – cannot arise from what is less perfect.[1]

Students reading Descartes for the first time often find that this is the point at which their patience finally snaps, and they cease giving him the benefit of the doubt. It is difficult not to sympathise with them: if there is one point at which – to the modern reader at least – the argument of the *Meditations* manifestly leaves the rails it is here.[2]

Several unflattering explanations can be given for Descartes's decision to proceed in this way, for example that he was incompetent in handling philosophical issues, or that he was less than wholly sincere when dealing with religious matters of any kind.[3] The former can be rejected without further delay; the latter is less easy to dispose of. Several of Descartes's contemporaries regarded him as an unbeliever who for reasons of prudence chose to give public allegiance to doctrines which in private he rejected.[4]

Fortunately the precise extent of Descartes's dissimulation need not be decided here. Metaphysical commitments like those expressed in the third Meditation can be found in other seventeenth-century philosophers whose sincerity there is no reason to doubt. In the *Cogitata Metaphysica* Spinoza offered a proof of the proposition that creatures are in God eminently by appealing to the principle that there must be at least as much perfection in the cause as there is in the effect, though it has to be said that from this thoroughly Cartesian principle he drew the very un-Cartesian conclusion that all the perfections of extension are in God (Spinoza 1972 I: 237). This particular argument does not reappear in the *Ethics*, but a metaphysical concept of perfection certainly does, and Spinoza made significant use of it.

No one has ever supposed that Spinoza was insincere about these matters, and there seems to be no need to assume that Descartes was either: I shall proceed on the assumption that he was not. If this is right then what we have is something very interesting, certainly to an intellectual historian, but I hope also to philosophers as well. The passages of the *Meditations* which we find alien and perhaps even only partially comprehensible are outcrops of an elaborate metaphysical system, now largely buried and requiring excavation if its structure is to be understood.

The system contains at least four elements:

1 There are degrees of reality, in the sense that some entities are more real, or more fully existent, than others.
2 The entities so distinguished can be ordered in a hierarchy, which has a highest (and in some versions also a lowest) member.
3 There is a fusion of the metaphysical and the ethical. Entities are graded in terms of *perfection*, and this is a concept that straddles what many modern philosophers would hold to be an absolutely fundamental distinction between facts and values.
4 Finally there is the principle that the cause must possess more perfection than its effect. I shall refer to this as the Causal Axiom.

Some comments on these seem appropriate:

1 The idea that there are degrees of reality, like so much else in this way of thinking, goes back to Plato; one thinks in particular of the analogy of the divided line in the *Republic*.
2 The idea that the degrees of reality can be ordered in a hierarchy could be seen merely as a corollary of the first principle, but is better listed separately. It is certainly possible for someone to hold a kind of unstructured pluralist view to the effect that there are many different ontologically distinct types of entity, but no ordering of any kind

among them; the Aristotelian categories, except substance, might be an example of this.

This doctrine can of course be expressed without any use of the word *hierarchy*. Though evidently Greek in its etymology, the word is not itself classical. It is ancient however: if not coined by Pseudo-Dionysius, the Christian Neoplatonist who assumed the identity of Dionysius the Areopagite around 500 AD, it was certainly established as a technical term by him, and it is no doubt because of his immense – if fraudulently acquired – influence, as well as the term's undoubted usefulness, that it passed into the common vocabulary of scholastic – and subsequently post-scholastic – philosophy.[5]

3 One of the metaphysical notions in Descartes that causes most trouble for the modern reader is that of perfection. Superficially there are no translation problems here: the English word 'perfection' translates the Latin *perfectio*, which in turn corresponds to the Greek τελειότης Understanding what these words once meant is however not quite so straightforward.

We may start with the Greek term. Τελειότης, like the adjective τέλειος from which it is derived, has connotations which the Latin *perfectio* lacks. Both words are derived from τέλος, *goal* or *end*, familiar to us from words like 'teleology'. The basic meaning is one of completion or fulfilment; a adult man can be described as τέλειος, meaning nothing more than that he is fully grown (Plato, *Laws*: 929C). None of these words is distinctively philosophical: all of them, even the abstract τελειότης, are to be found in the Greek New Testament (Heb. vi. 1, Col. iii. 14).

In Latin this semantic link is broken: there is no etymological connection between the word for end, *finis*, and *perfectio*, which comes from *perficere*, to complete, and ultimately from *facere*, to do or make. The severing of one connection merely results however in the establishment of another: the notion of perfection is now linked with that of making. Some of the implications of this new terminology are revealed in Aquinas' discussion of the perfections of God at the beginning of the *Summa Theologiae*. It might seem, he says, that God ought not to be described as perfect, because the perfect is, as it were, totally made (*quasi totaliter factum*), and God is not a being that is made at all. Clearly this is an argument founded on etymology of *perfectio*, and Aquinas saw no reason to be guided by such considerations. In philosophy the word needs to be given a different meaning: the perfect is *illud cui non deest esse in actu*, literally that which is lacking no being in act, and this description, unlike *quasi totaliter factum*, applies quite unproblematically to God.[6]

We have here in fact an interesting example of a not uncommon phenomenon. When the meaning of a term is fixed by its use in a tightly constructed intellectual system – whether scholastic theology, Newtonian mechanics, or whatever – its etymology largely ceases to matter; if the framework is loosened it regains, or at least can regain, its importance. An example is provided by the very different degree of attention that Locke and Spinoza give to the meanings of *sub* (under) and *stans* (standing) in their understanding of the term *substance*.

4 By the time that Descartes came to use it in *Meditations*, the Causal Axiom already had a long history. It appears as proposition 7 of Proclus' *Elements of Theology*: 'Every productive cause is superior to that which it produces', a proposition described by E. R. Dodds as 'the principle on which the whole structure of Neoplatonism is really founded' (Proclus 1963: 7, 193). Nothing can owe its existence to an entity inferior, or indeed merely equal, to itself. The causal order is therefore necessarily a hierarchy of decreasing perfection: 'in all that multiplies itself by procession, those terms which arise first are more perfect than the second, and these than the next order, and so throughout the series' (ibid., prop. 36).[7] If we were to imagine Proclus reading the *Meditations*, we could reasonably surmise that he would have been baffled by much that he found, but I am inclined to suspect that the argument for the existence of God in Meditation III would at any rate have appeared reassuringly familiar.

II

By no means all of Descartes's contemporaries found his line of thought persuasive. Hobbes found hierarchical metaphysics flatly unintelligible, and told Descartes so in an exchange that left each firmly persuaded of the metaphysical incompetence of the other. Hobbes announced his objection in a manner not calculated to calm Descartes's already mounting irritation:

> Moreover, M. Descartes should consider afresh what 'more reality' (*plus realitatis*) means. Does reality admit of more or less? Or does he think that one thing can be more of a thing than another? If so, he should consider how this can be explained to us with that degree of clarity that every demonstration calls for, and which he himself has employed elsewhere.
>
> (AT VII: 185)

Descartes's reply shows his annoyance:

> I have also made it quite clear how reality admits of more and less. A substance is more of a thing than a mode; if there are real qualities or incomplete substances, they are things to a greater extent than modes, but to a lesser extent than complete substances; and finally that if there is an infinite and independent substance, it is more of a thing than a finite and dependent substance. All this is completely self-evident [*Haecque omnia per se sunt notissima*].
>
> (AT VII: 185)

It was by no means completely self-evident to Hobbes.

This bad-tempered exchange between Hobbes and Descartes is interesting not only as a collision between two of the most self-confident, indeed arrogant, of the new philosophers, but also as one of the more easily visible manifestations of an intellectual revolution which has received relatively little attention.[8] A way of thinking that was dominant from late antiquity until the seventeenth century has since fallen into disuse. Many questions could be asked about this, but two in particular concern us. In the first place, how was it that this whole way of thinking became both unacceptable and unintelligible?[9] Secondly, what part if any did the natural sciences play in these events?

At first sight one might suppose the answer to the second question must be that the role of the natural sciences is unlikely to have been very great. The doctrine we are concerned with surely counts as a metaphysical one, if anything does, and quite manifestly is not open to empirical refutation. (One could add that many of its holders would not have regarded empirical refutation as having the requisite epistemic authority in any case.) Nevertheless the issue is well worth pursuing: metaphysical doctrines that become widely accepted and endure for centuries are seldom if ever adopted solely for their intrinsic merit. They appeal because they can be made use of, and are in turn discarded once they become useless or worse; in this respect they are like glue or cement, permeating structures composed of other materials and binding them together, or else failing to do so.

We can therefore quite reasonably presume that hierarchical metaphysics owed its enduring appeal to the fact that it stabilised and reinforced non-metaphysical doctrines. If we are seeking reasons for its decline we should certainly be prepared to look outside philosophy, though we should certainly still continue to look *inside* philosophy as well.

We may start however by looking outside. Even a cursory survey makes it clear that several quite distinct kinds of answer to the question of why hierarchical metaphysics declined have been proposed.

One is socio-political. Many twentieth-century thinkers, influenced directly or indirectly by Marx or Durkheim, are inclined to assume from the outset that religious and metaphysical doctrines are to a substantial extent projections of social structures. At its crudest – and it is generally the crude versions of any system that are the most influential ones – we have the familiar idea that grossly inegalitarian social arrangements are defended by being represented as natural – reflections of a supposed external reality which is in fact nothing more than a projection of the social relations themselves. During the early modern period European society underwent a change from a hierarchically ordered feudal society to a bourgeois capitalist one, and the dominant conceptual framework changed with it.

Any unease that one might feel about this picture need not arise from a belief that philosophical systems are devised and propagated in total isolation from society, but rather from a recognition that the connections are far more complex and less rigid than any simple 'ruling class: ruling ideas' model can allow. Consider for example the emergence of Neoplatonism in the third century AD. The Roman Empire in which Plotinus lived was a highly inegalitarian society, but not a strikingly *hierarchical* one as compared, for example, with medieval Europe. Medieval Europe on the other hand saw a systematic assault on hierarchical metaphysics in the form of Ockhamist nominalism. Some Marxist historians have seen a link between the revival of Neoplatonism in late fifteenth-century Florence and the rise of the Medici, the change being brought about by the decay of the republican institutions that had underpinned the old civic humanism. This is an interesting suggestion, but any *general* link between hierarchical metaphysics and political absolutism seems difficult to sustain.

Consider for example two philosophers from the seventeenth century. One – we may call him *A* – came from a clerical background, and was dependent for the whole of his life on the patronage of one of his country's wealthiest noble families. He was an unabashed supporter of political absolutism in its most extreme form, and when his country slid into civil war aligned himself unhesitatingly with the royalist side; albeit only with his pen, and at a safe distance, since he had fled abroad long before he himself could have been in any danger. He had a firm dislike of the republican values incorporated in the civic humanist tradition, which was exceeded only by his loathing of lower-class radicals and other disturbers of the traditional order.

Our second philosopher, *B*, came from one of the very few non-monarchical states in Europe, and the one with the most advanced, capitalist economy. In its political turmoils he was a firm and open supporter of the republican side, even at some risk to himself. By virtue of his own origin he was entirely detached from such remnants of the feudal hierarchy as remained in his country, and he earned the small income his modest tastes required from his labours as a skilled artisan, the very class in which (it is said) social and religious radicalism are most likely to be found.

Who is likely to be well-disposed to hierarchical metaphysics, *A* or *B*? If social milieu determined ideology the answer would surely have to be *A*. In the present context one can reasonably suspect that few if any readers would give this answer, partly because the strategy of the trick question is known to us all, but mainly because no one familiar with the seventeenth century would have much difficulty in penetrating the not very opaque disguises hiding Hobbes and Spinoza.

Hobbes's curt dismissal of Descartes' hierarchical metaphysics has already been described; it is not at all surprising that he had equally little time for more traditional notions of perfection. This is made clear in some passages in his most sustained attack on scholastic physics, the examination of Thomas White's *De Mundo*. Hobbes did not deny that the world can be described as perfect, but all this meant was that God had finished making it; it certainly did not imply that the world was best in any moral sense:

> The only person . . . who can deny that the world is made completely [*perfectissimum*] is he who at the same time wishes to deny that it was made as God wished it made. But could he have made a greater habitation, resplendent in more stars, or inhabited by more prudent and better animals, or resplendent in wiser or more pious persons who would have pleased God himself more? All this (unless the world is thought to be best on the grounds that God made it instead of another) is open to doubt.[10]
>
> (Hobbes 1976: 394, translation modified)

Such concepts as nobility also have no place in natural philosophy: 'Whether rest is nobler than motion is an absurd question, for nobility is the renown of men, [deriving] from lineage, riches, civil power, virtue and the like' (ibid.: 321). Hobbes did not insist that terms such as 'nobility' were to be eliminated from natural philosophy, but if they were to be retained they would need to be defined in a way that severed any link with human society – for example as special power of efficacy, or a potential to act (ibid.).

Hobbes never published his polemic against White, and Spinoza certainly never read it, but there can be little doubt that he would have agreed with Hobbes on one point at least. Nature is not to be regarded as imperfect because some aspects of it are uncongenial to human beings:

> The perfection of things is to be judged solely from their nature and power; things are not more or less perfect because they please or

> offend men's senses, or because they are of use to, or incompatible with, human nature.
>
> (*Ethics* I Appendix)

Spinoza's own position is however quite unlike Hobbes's in one crucial respect: while the anthropocentric elements of the traditional picture are dropped, the metaphysical elements are retained. There is a hierarchy of perfection, with God at its head: 'That effect is most perfect [*perfectissimus*] which is produced immediately by God, and the more something requires intermediate causes to produce it, the more imperfect it is' (*Ethics* I Appendix). Perfection can and should be defined in purely metaphysical terms: properly understood it is nothing other than reality.[11]

As in Proclus perfection is linked with agency; this is made clear in a much later passage in the *Ethics*, near the end of Part V: 'The more perfection a thing a thing has, the more it acts and the less it is acted on; and conversely, the more it acts, the more perfect it is' (*Ethics* V prop. 40). This is not a proposition that Hobbes would have been prepared to accept, and one may doubt whether he would even have found it intelligible.

III

If the fall of hierarchical metaphysics is not a mere by-product of social change, can it be explained in other ways? In particular, is there a link with the revolutionary changes that occurred in the natural sciences?

At first sight there are strong grounds for supposing that there is such a link. From late antiquity until the end of the sixteenth century the generally accepted model of the universe was Aristotelian, albeit modified and altered in various ways, most conspicuously by the introduction of epicyclic astronomy. The chief features of the Aristotelian cosmos was that it was finite and that it was *centrifocal*, to use David Furley's convenient expression (Furley 1989: vol. 1: 25). Contrary to what is often assumed this did not make the centre the most important point – indeed quite the opposite – but it did introduce an inherent *value gradient* into the universe: in contemplating the heavens one looked quite literally *upward*, and not merely *outward*, as we now do.[12] Because of its hierarchical organisation the Aristotelian-Ptolemaic cosmos was admirably suited to serve as a concrete, spatially extended manifestation or symbol of the metaphysical hierarchies of the intelligible world. The infinite universe of Bruno and Descartes entirely lacked this character, not because of its extent but because it was both homogeneous and isotropic; in such a universe there is no intrinsic difference between anywhere and anywhere else.

It is therefore unsurprising that some of the main proponents of the new astronomy had little sympathy with the old metaphysics. In Galileo's *Dialogue Concerning the Two Chief World Systems* Simplicio, the spokesman for the Aristotelians, began – following Aristotle's order in *De Caelo* – by arguing that the universe must have three dimensions because it is perfect. Salviati, representing Galileo himself, responded with unconcealed derision:

> I feel no compulsion to grant that the number three is a perfect number, nor that it has a faculty of conferring perfection upon its possessors. I do not even understand, let alone believe, that with respect to legs, for example, the number three is more perfect than four or two; neither do I conceive the number four to be any imperfection in the elements, nor that they would be more perfect if they were three. Therefore it would have been better for him [Aristotle] to leave these subtleties to the rhetoricians, and to prove his points by rigorous demonstrations such as are suitable to make in the demonstrative sciences.
>
> (Galileo 1967: 11)

When the notion of perfection was next mentioned it was again dealt with harshly, this time by the third participant, Sagredo:

> I cannot without great astonishment – I might say great insult to my intelligence – hear it attributed as a prime perfection and nobility of the natural and integral bodies of the universe that they are invariant, immutable, alterable etc., while on the other hand it is a great imperfection to be alterable, generable, mutable etc.
>
> (Ibid.: 58)

Like Hobbes, Galileo thought that notions such as perfection and imperfection, or nobility and baseness are entirely out of place in natural philosophy and have meaning only in relation to the circumstances of human life:[13]

> What greater stupidity can be imagined than that of calling jewels, silver, and gold 'precious', and earth and soil 'base'? People who do this ought to remember that if there were as great a scarcity of soil as of jewels or precious metals, there would not be a prince who would not spend a bushel of diamonds and rubies and a cartload of gold just to have enough earth to plant a jasmine in a little pot . . .
>
> (Ibid.: 59)

The kind of explanations Galileo preferred were of an entirely different character. While Simplicio argued that the heavenly bodies are all perfect, and consequently 'their shape is also perfect; that is to say, spherical',

Salviati replied by saying that the moon is (approximately) spherical because of the tendency of its parts towards their centre, not because the sphere is a perfect shape (ibid.: 84, 97).

The example of Galileo suggests an attractively simple solution to our problem: hierarchical metaphysics fell into disuse because hierarchical physics – the physics of the Aristotelian-Ptolemaic cosmos – was abandoned. Unfortunately things are not quite as simple as this. As the case of Descartes makes clear, there is no rigid connection between cosmology and metaphysics, such that anyone rejecting the old cosmology had necessarily to reject the old metaphysics, or vice versa. In natural philosophy Descartes and Hobbes were moderns: both were advocates of the new heliocentric cosmology and the mechanical world picture. Whatever the role of scientific theory-change in the fall of the hierarchical world picture, by itself it does not even begin to explain the deep metaphysical chasm – this seems hardly too strong a word – that manifests itself in the third set of *Objections and Replies*.

Quite a number of the real (or alleged) characteristics of Hobbes's thought can be eliminated fairly quickly as general causes of the decline of hierarchical metaphysics: in particular, materialism, atheism and naturalism. If the fall of hierarchical metaphysics had depended on the widespread acceptance of any of these, it would have occurred much later than it actually did – certainly not until the nineteenth century, possibly not until the twentieth. In fact it was rejected by most of Hobbes's successors.[14]

Though Locke cautiously endorsed a version of the Great Chain of Being (*An Essay concerning Human Understanding*, III. vi. 1–12) the *Essay* contains almost no trace of hierarchical metaphysics in any strict sense. The one partial exception occurs at the end of the proof of God's existence in Book IV, where we find the following passage:

> And whatsoever is first of all Things, must necessarily contain in it, and actually have, at least, all the Perfections that can ever after exist; nor can it ever give to another any perfection that it hath not, either actually in it self, or at least in a higher degree.
>
> (*Essay* IV. x. 10)

There are good reasons for supposing that this section of the *Essay* was written under the influence of Cudworth (Ayers 1991, vol. 2 170–6); certainly it seems to presuppose a type of metaphysics that Locke made little if any use of elsewhere.

Berkeley's attitude to materialism and atheism could hardly have been more unlike that of Hobbes, but at least in his youth he had no time for hierarchical metaphysics (the Neoplatonism of *Siris* is another matter). In his notebooks he bluntly rejected a key axiom of the older view: '*Nihil dat quod non habet* or the effect is contained in the Cause is an axiom I do not Understand or believe to be true.'[15] This is the Causal Axiom in its scholastic form, but Berkeley's objections were not merely to the schoolmen, as the very next entry in the notebook makes clear: 'Whoever shall cast his eyes on the writings of Old or New Philosophers & see the Noise is made about formal & objective Being Will etc.'[16] The grammar of this may be obscure, but the tone of disdain is not, and nor is the target (as the entries that follow make clear): the most conspicuous generator among the new philosophers of this particular variety of noise was Descartes.[17]

At this point it might appear that we have found the answer to our problem: the two parties to the dispute are neither the Ancients and Moderns, nor the Copernicans and anti-Copernicans, but rather our old friends the Rationalists and the Empiricists. Consider, after all, how the two parties seem to line up: on one side we have Descartes and Spinoza, and indeed Leibniz as well, while on the other we have Hobbes, Locke and Berkeley.[18]

At first sight this link with the rationalist/empiricist divide seems very plausible: the idea of perfection is a metaphysical idea, not obviously constructible out of any data provided by the senses, and is therefore presumably vulnerable to an empiricist critique. There are however some reasons for being cautious. *Nihil est in intellectu quod non prius erat in sensu* is after all good scholastic doctrine, accepted by Aquinas as well as Locke; nevertheless it is quite clear that Aquinas' acceptance of it in no way inhibited his employment of the metaphysics of perfection. The reason for this is not hard to find: the rules of this kind of concept-empiricism, if interpreted strictly, would rule out any kind of abstract thinking at all; if interpreted loosely – as they always are, even by Hume – they can be adjusted to let in pretty well anything one cares to choose.

The rejection of hierarchical metaphysics by the empiricists has therefore, I believe, to be explained on metaphysical rather than epistemological grounds.

Except for Berkeley, the empiricists did not see themselves as metaphysicians, but if one looks through their writings certain basic assumptions about the nature of reality do become apparent. One, explicitly asserted by all four, is arguably of crucial importance: this is the nominalist principle that everything that exists is an individual.[19] It was not a new idea in the seventeenth century, though it is likely that many of its proponents – Gassendi being an exception – were unaware of its earlier history. In some form or other it goes back to as far as hierarchical metaphysics itself, to the fourth century BC, though in this case it was to the opponents of Plato, such as Antisthenes, rather than to Plato himself. The first person to adopt the principle as a fundamental metaphysical axiom was William of Ockham, in the early fourteenth century.

The adoption of nominalism has profound implications for hierarchical metaphysics. In scholastic realism the hierarchy of genera and species – the Tree of Porphyry – was a metaphysical hierarchy of increasingly abstract and universal entities. For Ockham, as later for Locke, (Milton 1981: 128–45) it was purely a system of classification. The only real entities are those at the base: the individuals themselves.

Nominalism introduces therefore a kind of metaphysical egalitarianism: all individuals have the same basic ontological status. They are not of course equal in other respects – some are physically much *larger* than others for example, or are more important in other ways – but all are equally *real*: there is no hierarchy of existence.

In Ockham and his late-medieval followers the consequences of this metaphysical revolution were limited to philosophy, in the narrow sense of that word. Part of the cement that held together the Aristotelian world-picture had been removed, but the edifice as a whole was not disturbed. The universe retained its old hierarchical organisation, from the central earth to the outermost sphere of the *primum mobile*. The most important difference between the fourteenth-century nominalists and their seventeenth-century successors is to be found outside metaphysics: the former saw themselves as inhabiting a Ptolemaic universe, the latter a Copernican one.

It is clear that hierarchical doctrines in natural philosophy and in metaphysics are logically independent of one another: one can have both a physical hierarchy and a metaphysical hierarchy, as in Aquinas; a physical hierarchy without a metaphysical hierarchy, as in Ockham; a metaphysical hierarchy without a physical hierarchy, as in Descartes; and finally a rejection of both, as in Locke. The mixed options, if one may refer to them thus, are clearly possible, but history suggests that they have been less enduring than their rivals, and one reason for this is that they appear to be less intellectually stable.

In Descartes's thought the metaphysical theory of perfections is dangerously isolated. Historians of philosophy have been frequently been struck by the contrast between the centrality of God in Cartesian epistemology, and his near total absence from Cartesian physics. A similar divide appears here. In Descartes' metaphysics the notion of perfection is central and ineliminable; in his physics it is altogether absent.[20] Descartes saw no place for explanations of the sphericity of heavenly bodies in terms of the perfection of that shape; like Galileo he preferred to use efficient causes alone (*Principle of Philosophy*, iii. 61, iv. 19 ff.).

The effect of this revolution in physics was not to destroy hierarchical metaphysics, but to expose and thereby in the long run to undermine it.

Hierarchical metaphysics no longer had a secure place as a counterpart to hierarchical physics; it was on its own, standing or (increasingly) falling by what were perceived to be its own merits.

Notes

1 Translation taken from Descartes 1984, vol. 2: 28; also Descartes 1964–76 vol. 7: 40–1. Since the English translation also includes the Descartes 1964–76 page numbers in the margins, these alone have been given in the citations that follow, preceded by AT (Adam and Tanney, eds).

2 This is not of course an isolated passage: when in the *Second Replies* Descartes provided a formal exposition of his system *more geometrico* the same principle is listed as Axiom IV: 'Whatever reality or perfection there is in a thing is present either formally or eminently in its first and adequate cause' (AT VII: 166). It can also be found in the *Principles of Philosophy* (i.18).

3 Hobbes made both charges: according to John Aubrey he was sure that Descartes disbelieved in transubstantiation and defended it only to please the Jesuits; he thought Descartes would have done much better to stick to mathematics since 'his head did not lye for Philosophy' (Aubrey 1949: 94–5).

4 On the horrified reaction of Anna Maria van Schurman to some unusually indiscreet remarks made by Descartes when he called on her and found her reading the Old Testament in Hebrew, see AT IV: 700–1; also Verbeek 1992.

5 Lampe (1961) s. v. ἱεραρχία. The identity of Pseudo-Dionysius is unknown, and even his date uncertain, though at least some of his writings were in circulation by 532 AD (Pseudo-Dionysius1987: 13–14). His influence on Aquinas is described in O'Rourke 1992.

6 *Summa Theologiae*, I[a] q.4 a.1. For the two senses in which a created being can be perfect, see q.73 a.1.

7 This is Dodds's translation; the Greek is rather less ornate: τὰ πρῶτα τελειότερα τῶν δευτέρον ἐστι, καὶ τὰ δεύτερα τῶν μετ'αὐτὰ (the first are more perfect than the second, and the second than those after them).

8 Lovejoy (1936) is concerned more with changes in the world picture and their consequences than with their metaphysical underpinnings; it remains however essential reading. There are some interesting essays in Kuntz and Kuntz 1987.

9 This could seem like two questions, but I think that there is really only one; hierarchical metaphysics became unacceptable *because* it no longer seemed intelligible. I doubt that there has ever been anyone who thought that it made perfect sense to talk about perfections, degrees of reality etc., but nevertheless held that the world just is not like that.

10 The Latin original is in Hobbes 1973: 369.

11 'By reality and perfection I understand the same thing' (*Ethics*, II. def.6). What exactly this is meant to be a definition of is characteristically unclear.

12 In Dante the centre of the world is the lowest region of hell, occupied by Satan.

13 'It is scarcity and plenty that make the vulgar take things to be precious and worthless' (Galileo 1967: 59).

14 It was not rejected by everyone: a very traditional restatement can unsurprisingly be found in one of Hobbes's most determined critics, Ralph Cudworth. 'It being on the one hand, undenyably evident, that *Lesser Perfections* may

Descend from Greater, or at least from that which is *Absolutely Perfect* . . . but on the other hand utterly Impossible, that *Greater Perfections* and *Higher Degrees* of Being, should Rise and Ascend out of Lesser and Lower' (Cudworth 1678: 728).

15 *Philosophical Commentaries*, #780, Berkeley 1949–57, vol. 1. The literal sense of the Latin is: 'nothing gives which it does not have.'

16 Ibid., #781, quoted complete.

17 Especially ##782, 784–5.

18 Hume can, of course, be added to this group, though there is one ghostly remainder of the metaphysics of perfection in the *Treatise*, where Hume mentions what he describes as 'an establish'd maxim in both natural and moral philosophy, that an object which exists for any time in its full perfection without producing another, is not its sole cause'; he himself later endorsed the maxim when it reappeared as the last of his eight rules by which to judge of causes and effects (Hume 1978: 76, 174).

19 Hobbes 1991, chapter 5; Locke 1975, III.iii.1; Berkeley 1949–57, vol. 2: 192; Hume 1978, I.i.7: 19.

20 The same contrast is apparent also in Malebranche: he did not reject the hierarchical metaphysics completely, but retained it only for the order of perfections within the divine intellect (Malebranche 1980: 228–9, 617–8).

Bibliography

Aubrey, J. (1949) *Brief Lives*, ed. O. L. Dick, London: Secker and Warburg.

Ayers, M. (1991) *Locke*, London: Routledge.

Berkeley, G. (1949–57) *The Works of George Berkeley*, ed. A. A. Luce and T. E. Jessop, London: Nelson.

Cudworth, R. (1678) *The True Intellectual System of the Universe*, London: Richard Royston.

Descartes, R. (1964–76) *Oeuvres de Descartes*, ed. C. Adam and P. Tannery, Paris: J. Vrin.

Descartes, R. (1984) *The Philosophical Works of Descartes*, trans. J. Cottingham, R. Stoothoff and D. Murdoch, Cambridge: Cambridge University Press.

Furley, D. (1989) *The Greek Cosmologists*, Cambridge: Cambridge University Press.

Galileo (1967) *Dialogue Concerning the Two Chief World Systems*, trans. S. Drake, Berkeley: University of California Press.

Hobbes, T. (1991) *Leviathan,* ed. R. Tuck, Cambridge: Cambridge University Press.

Hobbes, T. (1973) *Critique du De Mundo de Thomas White*, ed. J. Jacquot and H. W. Jones, Paris: J. Vrin.

Hobbes, T. (1976) *Thomas White's De Mundo Examined*, trans. H. W. Jones, London: Bradford University Press.

Hume, D. (1978) *A Treatise of Human Nature*, ed. L. A. Selby-Bigge, rev. P. H. Nidditch, Oxford: Clarendon Press.

Kuntz, M. L. and Kuntz, P. G. (1987) *Jacob's Ladder and the Tree of Life: Concepts of Hierarchy and the Great Chain of Being*, New York: P. Lang.

Lampe, G. W. H. (1961) *A Patristic Greek Lexicon*, Oxford: Clarendon Press.

Locke, J. (1975) *An Essay Concerning Human Understanding*, ed. P. H. Nidditch, Oxford: Clarendon Press.

Lovejoy, A. O. (1936) *The Great Chain of Being* Cambridge, Mass.: Harvard University Press.

Malebranche, N. (1980) *The Search After Truth*, trans. T. M. Lennon and P. J. Olscamp, Columbus, Ohio: Ohio State University Press.

Milton, J. R. (1981) 'Locke and the nominalist tradition', *John Locke: Symposium Wolfenbüttel 1979*, ed. Reinhardt Brandt, Berlin: Walter de Gruyter.

O'Rourke, F. (1992) *Pseudo-Dionysius and the Metaphysics of Aquinas*, Leyden: E. J. Brill.

Proclus (1963) *The Elements of Theology*, ed. E. R. Dodds, 2nd edn, Oxford: Clarendon Press.

Pseudo-Dionysius (1987) *The Complete Works*, ed. C. Luibheid, London: SPCK.

Spinoza, B. (1972) *Spinoza Opera*, ed. C. Gebhardt, Heidelberg: Carl Winters.

Verbeek, T. (1992) *Descartes and the Dutch*, Carbondale, Ill.: South Illinois University Press.

6 Science in the service of life

Nietzsche's objectivism

Aaron Ridley

Introduction

Nietzsche's thoughts about science, as about everything else, are complex and involved. They are also quite readily intelligible.[1] But they cannot be understood unless certain preconceptions about Nietzsche are discarded or put on hold. He is, for instance, widely believed to be an irrationalist. But in fact he only rejects a certain conception of rationality, a Kantian conception, and on grounds which Nietzsche, perhaps not unreasonably, regards as rational. Any attempt to reject rationality as such would have struck him as absurd, and worse. Nietzsche is also widely thought to be a relativist of an extreme and especially pernicious sort. But, again, it is only a particular conception of realism that he rejects, and for reasons that he certainly does not regard as relative. If one persists, however, in mistaking Nietzsche for an irrationalist and a relativist then it is all too easy to get him wrong on science as well. He is not, if one reads him at all carefully, opposed to science (although he is opposed to scientism). But the metaphysically realist conception of science that he singles out for attack is sufficiently popular for opposition to it to be mistaken for opposition to science as such. The truth, as I will try to show, is that Nietzsche is committed to a positive, rational conception of scientific enquiry, and to the rigorous standards of objectivity that it presupposes.[2] This may make him a less excitingly offbeat figure than he is sometimes taken to be. But it has the compensation, I think, of getting him right; and it also makes him, in a sense having nothing to do with eccentricity, more interesting.

Perspectivism

The charges of irrationalism and relativism levelled against Nietzsche can be defused through an exploration of the epistemological and metaphysical context within which science, in his view, is to be understood. I shall begin, then, with Nietzsche's famous, not to say notorious, statement of the epistemological-cum-metaphysical position known as perspectivism:

> let us be on guard against the dangerous old conceptual fiction that posited a 'pure, will-less, painless, timeless knowing subject'; let us guard against the snares of such contradictory concepts as 'pure reason,' 'absolute spirituality,' 'knowledge in itself': these always demand that we should think of an eye that is completely unthinkable, an eye turned in no particular direction, in which the active and interpreting forces, through which alone seeing becomes seeing *something*, are supposed to be lacking; these always demand of the eye an absurdity and a nonsense. There is *only* a perspective seeing, *only* a perspective 'knowing'; and the *more* affects we allow to speak about one thing, the *more* eyes, different eyes, we can use to observe one thing, the more complete will our 'concept' of this thing, our 'objectivity,' be.
>
> (Nietzsche 1967, Essay III, section 12)

Nietzsche is here summarising three complementary accounts: one of the nature of knowers, one of the nature of knowledge and one of the nature of the objects of knowledge.

Knowers, on Nietzsche's account, are thoroughly embedded in the world and in history. To think of them as Kant thought of them, as 'pure, will-less, painless, timeless knowing' subjects, is to extrude from them precisely those embedded and embodied features that make knowing possible at all. Suppose, Nietzsche suggests, that we were able 'to eliminate the will altogether, to suspend each and every affect . . . – what would that mean but to *castrate* the intellect?' (Nietzsche 1967 Essay III, section 12). The intellect would be castrated, he holds, because among the 'affects' to be suspended would be the desire to know ('the will to truth' (Nietzsche 1967: III.27)) and the desire to produce rationally acceptable explanations of the phenomena we know about (see Nietzsche 1967: III. 12).[3] To suspend these would be to leave behind only the 'nonsensical absurdity' of 'contemplation without interest', i.e. of contemplation somehow conducted in the absence even of our cognitive interests (in things like simplicity, explanatory power, etc.), let alone of those other interests (in things like convenience, survival, etc.) that give us reasons for wanting to know anything in the first place (Clark 1990: 48). The knower, then, on Nietzsche's account, is situated, not in some abstract state of pure, disembedded knowing, but in the interstices of those patterns of interest and desire that he calls 'system[s] of purposes' (Nietzsche 1967: II. 12). This is what it is for a knower to have a 'perspective'; and perspective is an essential prerequisite of knowing.

Nietzsche's account of knowledge dovetails with his account of knowers. Negatively, knowledge is not 'knowledge in itself'; i.e. it is not the sort of information that might be garnered by an impossibly abstract, disembedded knower. Rather, because generated by knowers whose perspectives are defined by particular systems of purposes, knowledge is always a function of the interests immanent to those systems. So, for instance, the standards of rational acceptability appropriate to one kind of enquiry may not be appropriate to others. What counts as evidence or as an argument in mathematics, say, may be quite different from what counts as evidence or as an argument in ethics: the cognitive interests at issue are different. Therefore there is no such thing as perspectiveless knowledge; or, to put the point another way, there is no such thing as *the* truth (or Truth), even if, relative to a given system of purposes, some things are certainly true and others certainly false.

Taken together, these conceptions of knowers and of knowledge entail a conception of the objects of knowledge, a conception apparent in Nietzsche's remark that it is only through perspective that 'seeing something becomes seeing *something'*. The point here is that because all of our encounters with objects take place, and must take place, in the context of our interests and purposes, we encounter objects in the terms set by those interests and purposes. If we had no interests or purposes, no perspective, objects would be encountered in no terms at all, and so would not, in any intelligible sense, be encountered. It is for this reason that Nietzsche is so scathing about the Kantian 'thing-in-itself', the elusive essence of an object that supposedly underlies its 'mere' appearance.[4] '[W]hat could I say', Nietzsche asks, 'about any essence except to name the attributes of its appearance! Certainly' an appearance is 'not a mask that one could place on an unknown X or remove from it!' (Nietzsche 1974 sect. 54). Thus, in Nietzsche's view, the very idea of a 'thing-in-itself' – of an essence, an 'unknown X' – to which appearances somehow accrue, is unintelligible, 'a *contradictio in adjecto*' (Nietzsche 1966 sect. 16), since the subtraction from a thing of all of its possible appearances simply constitutes the subtraction of the thing itself. But once we 'have abolished the real world' – i.e. the world of things-in-themselves – 'what world is left? the apparent world perhaps? But no! *with the real world we have also abolished the apparent world*!' (Nietzsche 1968). Thus the objects of knowledge are neither Real, in the way that things-in-themselves were supposed to be real, nor are they merely appearances. Instead, they are just objects: things singled out as such as a result of the ways in which they impinge on various systems of purposes, and conceived in the terms set by those systems. Which is why Nietzsche insists that 'the *more* eyes, different eyes, we can use to observe one thing, the more complete will our 'concept' of this thing, our 'objectivity,' be.'

The position that Nietzsche opposes, then, is a Kantian, metaphysically realist one, according to which true knowledge is a grasp of things as they are in themselves, a grasp possible only for a knower who has somehow transcended any and every possible system of purposes. In Nietzsche's view, this position, apart from being contradictory, also exhibits a 'horror . . . of reason itself' (Nietzsche 1967: III. 28). Indeed he goes further:

> To renounce belief in one's ego, to deny one's own 'reality' – what a triumph! not merely over the senses, over appearance, but a much higher kind of triumph, a violation and cruelty against *reason* – a voluptuous pleasure that reaches its height when . . . reason declares: '*there* is a realm of truth and being, but reason is *excluded* from it!' . . . [Thus] the Kantian concept of the 'intelligible character of things' . . . signifies . . . that things are so constituted that the intellect comprehends just enough of them to know that for the intellect they are – *utterly incomprehensible*.
>
> (Nietzsche 1967: III.12)

Nietzsche's own position, by contrast, is a version of what Hilary Putnam calls 'internal realism': knowers are embedded in systems of purposes and so have interests, including cognitive interests; the concept of knowledge is a function of those purposes and interests, and so itself is embedded; and the objects of knowledge are those features of the world that achieve salience in the context of, and hence are conceived in terms of, the systems of purposes that knowers inhabit.[5] This position is not irrationalist. To reject the Kantian idea of a reason so 'pure' that it transcends all possible perspectives is not to reject reason itself: on the contrary, it is to locate reason in the real, embodied world in which it actually does its work. Nor is this position relativist. To deny that there are truths about things as they are in themselves is simply, in Nietzsche's view, to unmask a contradiction; and to insist that knowledge is perspectival is merely to say that what *counts* as true or as evidence or as an argument is not independent of the systems of purposes within which knowledge is sought. (Nor, I suppose it is worth spelling out, is Nietzsche a relativist on the grounds that perspectivism, if true, must itself be merely a perspective. His point, rather, is that the denial of perspectivism, as exemplified by the Kantian conception, is simply unintelligible. Maudemarie Clark puts it nicely when she observes that just as 'creative power is not limited by the inability to make a square triangle,' so 'cognitive power is not limited by the inability to have nonperspectival knowledge' (Clark 1990: 134).)

Value and truth

Because, on Nietzsche's analysis, there is no such thing as 'knowledge in itself', there is no such thing, either, as enquiry in itself. Every form of enquiry, including scientific enquiry, must be understood in terms of the system of purposes by which it is motivated and from which it draws its special rationale. '[T]here is no such thing,' as he puts it, 'as science "without any presuppositions"'; for 'a philosophy, a 'faith,' must always be there first of all, so that science can acquire from it a direction, a meaning . . . a *right* to exist' (Nietzsche 1967: III.24). Nietzsche is pretty confident that he knows what this faith is, at least as it animates the best sort of scientist. It is the 'unconditional will to truth.' What 'constrains *these* men,' he says, 'is the faith in a *metaphysical* value, the absolute value of *truth*, sanctioned and guaranteed by this ideal alone' (ibid.). Nietzsche's scientist, then, is motivated by the conviction that truth is 'inestimable and cannot be criticized' (Nietzsche 1967: III.25), that 'truth is *divine*' (Nietzsche 1967: III.11); which is to say, the scientist is committed to the view that there is a single, final Truth to be got at, a perspectivelessly true take on the world as it really is (in itself). This, of course, is an expression of the sort of metaphysical realism that Nietzsche, as we have seen, rejects. But it is also an expression of something deeper. Nietzsche puts it like this:

> The truthful man, in the audacious and ultimate sense presupposed by faith in science, *thereby affirms another world* than that of life, nature, and history; and insofar as he affirms this 'other world,' does this not mean that he has to deny its antithesis, this world, *our* world? . . . It is still a *metaphysical faith* that underlies our faith in science.
>
> (Nietzsche 1974: 344)

Thus, just as Kant sought to 'triumph' over his own 'ego' and his own 'reality' when he invented 'a realm of truth and being,' so the scientist seeks to triumph over 'life, nature, and history' by placing truth *outside* the world of purposes and interests that we actually inhabit (i.e. by turning truth into Truth). Both attempts, in Nietzsche's terms, are expressions of the *ascetic ideal*: the determination to locate the value and the meaning of life elsewhere, in a disembedded, ahistorical 'beyond,' a fantasy-land of things as they are in themselves. The ascetic ideal 'juxtaposes' life '(along with what pertains to it: "nature," "world," the whole sphere of becoming and transitoriness) with a quite different mode of existence which it opposes and excludes, *unless* it turn against itself, *deny itself*'; thus, from the perspective of the ascetic ideal, 'life counts as' merely 'a wrong road . . . a mistake' (Nietzsche 1967: III.11).

The ascetic ideal owes its power to the fact that it is, as Nietzsche puts it, 'a closed system of will, goal, and interpretation' (ibid.: III.23). It posits a goal: the 'beyond.' It assigns to that goal unconditional metaphysical value, thereby rendering the will to pursue it self-justifying. The pursuit of the goal then takes the form of interpretations which confirm the unconditional value of the 'beyond' by denigrating 'life, nature, and history,' and whatever else pertains to 'this world, *our* world.' In this much, then, the scientist's 'unconditional will to truth' is part of the package of values, ranging from Platonism and Christianity through to Kantianism and Wagnerism, that the mature Nietzsche regards as a treason against life. But science occupies a special position in this spectrum. Its faith in truth has the effect of undermining other constituents of the ascetic package, most notably religion, and ultimately of undermining the ideal itself. First science 'forbids itself *the lie involved in belief in God*.' But the lies it forbids itself can't stop there. Eventually, Nietzsche says, after it 'has drawn one inference after another, it must end by drawing its *most striking inference*, its inference *against* itself' (ibid.: III.27); namely, that the assignation of transcendental status to truth is just as contradictory as the assignation of that status to God. If one takes the truth seriously enough, in other words, one will eventually forbid oneself the lie involved in the belief in Truth (i.e. in 'knowledge in itself' of things in themselves). One will, in short, become a perspectivist; one will situate the pursuit of truth (not Truth) in the context of the immanent interests and purposes that actually give rise to it.

But, still, there can be no such thing as 'science 'without any presuppositions"; science is not 'self-reliant'; it requires 'an ideal of value, a value-creating power', for 'it never creates values' itself (Nietzsche 1967: III.25). For the reasons we have seen, it can no longer make the presuppositions it has made until now. It cannot posit Truth as an unconditionally valuable goal whose pursuit is metaphysically self-justifying. Nor, given its final inference, can it instate anything else as that, either. With the demise of the transcendental authority accorded to it by the ascetic ideal, 'Science henceforth *requires* justification' (ibid.: III.24) ; and that justification is going to have to be provided in terms of immanent, rather than other-worldly, systems of purposes. In Max Weber's words, we need new reasons to think 'that what is produced by scientific work' is '*important* in the sense of 'being worth knowing.' And it is obvious that all our problems lie here, for this presupposition cannot be proved by scientific means' (Weber 1989: 18).

Scientism

Such, in outline, is Nietzsche's general account of science. We are now in a position to make sense of some of his more specific remarks.

One of the charges Nietzsche makes against Kant is that he attempted to triumph 'over the senses,' the senses of course being too obviously this-worldly and perspectival to be fit for the pursuit of Truth. In light of this, one would expect Nietzsche to be pretty keen on the purely empirical aspects of science. And so he is:

> What magnificent instruments of observation we possess in our senses! . . . Today we possess science precisely to the extent that we have decided to *accept* the testimony of the senses – to the extent to which we sharpen them further, arm them, and have learned to think them through. The rest is miscarriage and not-yet-science.
>
> ('Reason and philosophy', Nietzsche 1968)

To '*accept* the testimony of the senses' is clearly to accept the senses as sources of reliable information about the actually embodied world, about 'the whole sphere of becoming and transitoriness.' To 'think' the senses 'through' is to realise that the information they provide is not about things as they are in themselves, but about those real objects that impinge upon observers and their systems of purposes. This detranscendentalised conception of science is entirely to Nietzsche's liking; and it is in the spirit of it that he remarks that 'physics, too, is only an interpretation and exegesis of the world (to suit us, if I may say so!) and *not* a world-explanation' (Nietzsche 1966: 14). To describe physics as an 'interpretation and exegesis' is to highlight its perspectival character; to say that its interpretation is 'to suit us' is to underline the fact that knowing is always rooted in interests, is always part of a system of purposes; and to deny that it is 'a world-explanation' is to deny that the truths of physics are disembeddedly True of the world as it is in itself.

Problems arise only when science oversteps its limits and claims transcendental authority for its findings. This is not usually done explicitly. It is usually done by the expedient of denying that the scientific view of the world is, itself, a perspective. In this it shows its deep affinity to the ascetic ideal, which:

> rejects, denies, affirms, and sanctions solely from the point of view of *its* interpretation . . . ; it submits to no other power . . . – it believes that no power exists on earth that does not first have to receive a meaning,.. a value,.. as a way and means to *its* goal, to *one* goal.
>
> (Nietzsche 1967: III.23)

So successful can such metaphysical *coups* be that they become all but invisible (Nietzsche 1967: I.7), so that their interpretations come to be taken as simple reports of the way things are, independently of any and every set of interests or purposes. Their interpretations come to be seen as True. Once this happens, alternative perspectives are rendered marginal at best, and more likely unimaginable. When the interpretations at issue are scientific interpretations, the result is scientism: the false and ascetic belief that the scientific perspective enjoys a uniquely privileged mode of access to reality (or Reality), i.e. the false and ascetic belief that science does, indeed, have transcendental authority. Nietzsche is acutely aware of the scientism of his contemporaries, and had he had us for his contemporaries his awareness would have been acuter still. What alarms him is the way in which the conception of objects integral to the scientific perspective has come to be seen as *the* real and true way of conceiving of objects. Take the following passage: as the result of a 'passive, automatic, reflexive, molecular'picture of the world, the:

> absolute fortuitousness, even the mechanistic senselessness of all events . . . has permeated the realm of the spirit... to such a degree that today it has forced its way . . . into the strictest, apparently most objective sciences . . .– to the detriment of life, as goes without saying, since it has robbed it of a fundamental concept, that of *activity* . . . [replacing it with] a mere reactivity.
>
> (Nietzsche 1967: II.12)

Nietzsche's point is not that 'activity' is somehow a truer property of life as it is in itself than 'mere reactivity,' but that the mechanistic conception prevents us from thinking about life from perspectives that would make the concept of 'activity' central; and for some purposes, he believes (his whole philosophy is testament to the fact), such perspectives may serve our interests better, or better serve our more important interests, than the scientific one does.

Or consider his well-known aversion to Darwinism. Darwinism, he says, because its proponents ('the English') know 'the difficulties of survival all too well at first-hand,' espouses an 'incomprehensibly one-sided doctrine of the 'struggle for existence'.' But a true 'natural scientist' should recognise that:

> in nature it is not conditions of distress that are *dominant* but overflow and squandering, even to the point of absurdity. The struggle for existence is only an *exception*, a temporary restriction of the will to life. The great and small struggle always revolves around superiority, around growth and expansion, around power.
>
> (Nietzsche 1974: 349)

Nietzsche's problem here is, in effect, the same as before. The conception of life that Darwinism presupposes and promotes is partial, 'one-sided'. It focuses solely on those aspects of life that can be understood as the products of mechanism, as the results of the shaping effects of the environment *on* life. What Nietzsche calls 'the will to life' can find no place in this conception. His point, then, is not that Darwinism is false. In the context of the system of purposes constitutive of science, it may very well be true; indeed, Darwinism may well offer the best possible account of life so conceived. Nietzsche's point, rather, is that if one becomes persuaded that the conception of life implicit in Darwinism is *the* conception of life, that the Darwinian conception exhausts the phenomenon, one deprives oneself of the means to think about life in other terms; and for some purposes, for instance ethical purposes, it may be altogether more valuable to think of life as, for example, 'spontaneous, aggressive, expansive, form-giving' (Nietzsche 1967: II.12) than to think of it as the product and expression of mere inanimate mechanism. Scientism is a variety of perspective-blindness, then: to be in its grip is to mistake the perspective of science for the view from nowhere, to confuse objective scientific truth with truth and objectivity as such. Which is why Nietzsche insists that objectivity as such requires 'the ability *to control* one's Pro and Con and to dispose of them,' that is, the ability to adopt and to suspend points of view, 'so that one knows how to employ a *variety* of perspectives and affective interpretations in the service of knowledge' (ibid.: III.12).

But of course science and scientism are not the same thing. To insist, against scientism, that the scientific view is merely one perspective among many is not to criticise the perspective of science; and Nietzsche, indeed, nowhere does criticise that perspective. It is clear to him, as to anyone sensible, that the methods and presuppositions of science serve certain of our interests much better than the alternatives would. As an explanation of the heavens in their causal aspect, astronomy is infinitely superior to astrology. As an explanation of the weather in its mechanical aspect, meteorology is almost certainly better than demonology. As an explanation of the body in its fallible aspect, medical science is at least quite arguably an advance on the theory of the humours. Nietzsche's point is only that the reverse is sometimes also true. As an explanation of political action, for instance, Machiavelli is superior to rational choice theory; of human behaviour, folk psychology to the other sort; of music, aesthetics to acoustics. One has to be clear, that is, which *values* are at issue before it is at all appropriate to conclude that science is where to go for the answers. Science may have nothing or nothing worth knowing to say about a given phenomenon in the context of a given system of purposes (for instance, about music as worth listening to). Indeed, science may even have things to say that are worth not knowing. A perfectly plausible case can be made, for example, for the claim that opinion polls ought not to be taken in the lead-up to general elections. This is not because they are not true: they may very well provide a scientifically objective measure of voting intentions. It is, rather, because, in the context of certain political values, the effects of that information on the political process (tactical voting, policy-tailoring, etc.) may be politically pernicious. It may serve our citizenly interests better, in other words, if such things are not known. And this, in effect, is the point behind another well-known claim of Nietzsche's, that 'the value of truth must for once be... *called into question*' (Nietzsche 1967: III.24). The point is not that truth may be unimportant. Truth is clearly important, and Nietzsche never denies or doubts it. The point is rather that the mere fact that something is true, for instance that x per cent of the population intend to vote for y, does not by itself constitute a reason or justification for finding it out or for knowing it. Truth, in that sense, is not of inestimable value, is not '*divine*'.

Conclusion

Nietzsche had been bothered by questions of truth, value and knowledge since the very beginning of his career. In *Untimely Meditations*, for instance, we find him worrying, not about science, but about history:

> We need history, certainly, but we need it . . . for the sake of life and action, not so as to turn comfortably away from life and action, let alone for the purpose of extenuating the self-seeking life and the base and cowardly action. We want to serve history only to the extent that history serves life: for it is possible to value the study of history to such a degree that life becomes stunted and degenerate – a phenomenon we are now forced to acknowledge . . . in the face of certain striking symptoms of our age.
>
> (Nietzsche 1983: 59 (foreword))

At this point in his development, Nietzsche's metaphysical and epistemological views were still largely the product of Schopenhauer's influence. He decided, as a result, that truly objective history 'history in the service of life' would have to be 'supra-historical' and would have to speak of '*eternal verities*' (Nietzsche 1983: section 10). The mature Nietzsche, of course, can accept no such position. But the project that Nietzsche the Schopenhauerian was pursuing – to ask how, if knowledge isn't self-justifying, knowledge is to be justified – and the terms in which it was couched – life and the service of life – are exactly the project and terms we find in the later Nietzsche, even if his late answer is more or less the opposite of his early one. The young Nietzsche is still caught up in the problem he is trying to diagnose. He wants to make the value of a certain kind of knowledge contingent on its service to life; but he also wants to attribute other-worldly status to that very same kind of knowledge (by turning historical knowledge into supra-historical Truth). For the mature Nietzsche, by contrast, the demands of 'life' and of intelligible metaphysics both require that nothing, least of all knowledge, be accorded that status. And hence the objectivism of his later reflections upon science: objectivity consists in knowing why and if what one knows or wants to find out is worth knowing. Or, to put it another way, it consists in acknowledging that knowledge is perspectival, that perspectives express interests, and so that knowledge can be, and can fail to be, in the service of life. Science fails to be in the service of life when its perspective is taken to be the only perspective there is. Science so conceived is a version of the ascetic ideal: it betrays life by denying 'this world, *our* world.' Or, to put it another way, science so conceived betrays life by preventing us from thinking about ourselves and the world in ways that may, in fact, be in our interests. This, in a nutshell, is what Nietzsche's later discussions of knowledge, and especially of scientific knowledge, come to; and if his ideas about it no longer seem so tremendously radical, that may be because they have had some of their due effect. Through the writings of Weber and Wittgenstein, of Kuhn and Foucault, we have become familiar with them, or with their general tenor. But they were startling ideas when he first had them; and their implications, I suspect, have still to be fully thought through.

Notes

1 The term 'science' is tricky in Nietzsche. It is used to translate the German *Wissenchaft*, a word connoting not only what in English would be called science, but knowledge and wisdom in general. In this essay I concentrate on the things that Nietzsche says that either apply to knowledge as such (and so to 'science' as a sort of knowledge) or else apply more specifically to what, in English, would be called science.

2 I have made this case from a different angle elsewhere, in Ridley 1998 chapter 5. My account here (as there) is heavily indebted to Clark 1990 chapters 1–4, and to Owen 1995 chapter 2.

3 The 'will to truth' can itself be a bar to knowledge, but only when it is unconditional.

4 See for example, *Beyond Good and Evil* (Nietzsche 1966), section 2.

5 See, for instance, 'A defense of internal realism' in Putnam 1990. For a useful discussion that highlights slightly different aspects of perspectivism, see Leiter 1994.

Bibliography

Clark, M. (1990) *Nietzsche on Truth and Philosophy,* Cambridge: Cambridge University Press.

Leiter, B. (1994) 'Nietzsche on truth', in R. Schacht (ed.) *Nietzsche, Genealogy, Morality*, Berkeley: University of California Press.

Nietzsche, F. W. (1966) *Beyond Good and Evil,* trans. W. Kaufmann, New York: Vintage.

Nietzsche, F. W. (1967) *On the Genealogy of Morals,* trans. W. Kaufmann and R. J. Hollingdale, New York: Vintage.

Nietzsche, F. W. (1968) *Twilight of the Idols,* trans. R. J. Hollingdale, London: Penguin.

Nietzsche, F. W. (1974) *The Gay Science*, trans. W. Kaufmann, New York: Vintage.

Nietzsche, F. W. (1983) 'On the uses and disadvantages of history for life', in *Untimely Meditations* trans. R. J. Hollingdale, Cambridge: Cambridge University Press.

Owen, D. (1995) *Nietzsche, Politics and Modernity,* London: Sage.

Putnam, H. (1990), *Realism with a Human Face*, ed. J. Conant, Cambridge: Harvard University Press, 30–42.

Ridley, A. (1998) *Nietzsche's Conscience: Six Character Studies from the Genealogy*, Ithaca: Cornell University Press.

Weber, M. (1989) 'Science as a vocation', in P. Lassman and I. Velody (eds) *Max Weber's 'Science as a Vocation',* London: Unwin Hyman.

7 Pragmatism

Commonsense and the limits of science

Christopher Hookway

Pragmatism and science: norms of inquiry

A recent book by Howard Mounce is called *The Two Pragmatisms* (Mounce 1997). The title refers to the positions of two of the classical American pragmatists of the late nineteenth century, Charles S. Peirce and William James; and Mounce argues for the greater originality and contemporary interest of the latter. In this he follows Richard Rorty who once remarked that Peirce 'gave only a name to pragmatism'(Rorty 1982: 161); and, although others have different views of who is villain and who is hero, such a distinction is commonly drawn. It rests upon a contrast in answers that can be given to the question that concerns us here. Peirce is a scientifically minded systematic philosopher who thinks that scientific knowledge is the only real knowledge, while James describes a pluralistic universe in which different bodies of belief and different styles of inquiry can answer to different cognitive and affective needs.

The same contrast is evident when contemporary philosophers are described as pragmatists, or when they try to enrol themselves under that banner. Rorty emphasises that there are no substantive epistemic differences between scientific knowledge and (say) ethical knowledge, and takes from James the view that 'True' is just a linguistic device for expressing our approval of beliefs (Rorty 1982: xii). He would presumably deny that we can point to any fundamental standards of rationality which attest to science's special status: its prestige derives solely from its *de facto* success. Others, like Patricia Churchland, see themselves as heirs to an anti-foundationalist tendency in pragmatism when they use the term to describe the use of scientific information to propose revisions to our epistemological categories and our everyday conceptions of mind. John Dewey's readiness to describe his pragmatism as a form of 'naturalism' encourages this perspective.

This paper will focus on the classical pragmatists: on Charles S. Peirce,

William James and John Dewey. I suspect that the 'two pragmatisms' picture overstates the differences between them and that elements of both pictures are present in each thinker. If this is correct, it can help us to describe and understand some tensions within the thinking of each. It will also be useful in helping us to identify the distinctive understandings of 'science' that they bring to their work. In the following section, I will identify some passages which support the view that they have dramatically contrasting attitudes towards science. Before doing so, we should note an important notion in the work of all pragmatist philosophers: *inquiry*.

Peirce introduced a theory of inquiry in his paper 'The fixation of belief', and it is arguable that adopting this as a framework for thought about epistemic matters is characteristic of pragmatism. Inquiry is an activity of problem solving, one which always occurs within a context, against a background of beliefs that currently hold firm. For Peirce, it begins (most commonly) when perceptual surprise disturbs a stable body of belief, introducing a doubt which needs to be removed. Once that doubt has been replaced by a stable (albeit fallible) belief, we are content: the goal of inquiry is stable belief. This 'pattern of inquiry' is developed by Dewey (1938). His view of the origins and upshots of inquiry is less propositional, more holistic. We are motivated to inquire by finding ourselves in an 'indeterminate situation': in some fashion, our expectations, experience, habits of action and our goals and surroundings fail to fit together. Distilling our uncertainty into a concrete 'problem' is a cognitive achievement, the first stage of investigation; and the first source of possible error, for we can institute a problem whose solution in fact fails to remove the initial indeterminacy. In that case, inquiry comes to a close not when we have arrived at a belief which answers some initial *question*, but rather in the fact that our 'situation' has become determinate again. Both Peirce and Dewey insist that such a pattern is common to everyday inquiries and those found in the sciences; and the latter, in particular, insists that a simpler version of his pattern is found in the animal world too. Although James has little to say about the structure of inquiry, the view of the growth of knowledge that emerges from the Peirce-Dewey picture is found in his work too.

What distinguishes 'scientific' inquiries from those of other types is an issue we must discuss more fully later. For the present, it is important to note that the pragmatists seem to offer different answers to this question. For Dewey, 'science' is distinguished from, for example, 'common sense' investigations by its subject matter: it is concerned with abstract system of relations. Methodological differences between the two will then be consequential upon these differences in subject matter. Peirce, as we shall see, appears to draw the distinction differently: scientific inquiries are distinguished, not by their subject matter, but by the 'spirit' with which they are carried out, and by the 'methods' employed.

It will be helpful to say something about the word 'pragmatism'. Uncontroversially, pragmatists deny the sharp separation of theory and practice which pervades some earlier empiricist philosophy: we cannot study epistemic rationality without taking account of the fact that inquirers are agents whose assessment of their beliefs is embedded in a network of goals, emotions, intentions and interests. This observation can interact with these comments about 'inquiry' in two different ways. One view would be that we cannot study the evaluations made when we carry out inquiries unless we take account of the fact that theoretical inquiries are often subordinate to the achievement of practical goals: we need information that will help us to achieve our goals or solve other practical matters. The indeterminacies in our situations may be produced by (or salient because of) our values and desires. For our purposes a second point is more important. Inquiry itself is an activity, driven by the internal goal of settling belief or rendering an indeterminate situation determinate. Moreover, both Peirce and Dewey emphasis that their view of inquiry is an *experimentalist* one. When Peirce takes account of how experience can guide inquiry, he is primarily concerned with the ways in which our activities can elicit or produce experiences: he describes his pragmatism as a 'laboratory philosophy'. And Dewey emphasises the role that manipulating and interfering with the situation can have in the process of rendering it determinate. One way to understand the idea that there are two distinct strains within pragmatist thought emerges here. Peirce's work takes very seriously the second way in the which practical and theoretical rationality interact but he often appears to be suspicious of inquiries and systems of classification that are subordinated to our non-scientific purposes. The second strain, associated with Williams James, and celebrated by contemporary pragmatists such as Rorty, emphasises the role of practical goals and interests in guiding and shaping our knowledge (see Elgin 1997, chapter XI)

Peirce used the word 'pragmatism' to refer to a distinctive rule for clarifying the meanings of concepts and hypotheses: we achieve clarity by reflecting on the experiential consequences we would expect our actions to have if the hypothesis were true or if the concept applied to some accessible object. Concepts mediate connections between actions or interferences on the one hand and experiences on the other. Of course this is relevant to how our knowledge is applied in achieving our practical goals, but it is reasonable to say that, for Peirce, these connections are of primary interest for their role *within* inquiry. We exploit these connections in making observations and designing experiments; they help us to arrive at satisfactory solutions to our problems. James used the word with a broader meaning. Sometimes he says that pragmatism is a 'theory of truth'; alternatively, it is described as a philosophical method or attitude, one that helps us to dismiss traditional philosophical problems by showing that no 'practical' difference turns on the acceptance of one solution rather than another.[1]

For all of our pragmatists, the fundamental epistemological (or 'logical') questions concern the norms we should follow in trying to order and plan our inquiries and deliberations. If we allow that the sort of pattern described above is common to *all* inquiries, then questions about the limits of science concern the nature of these norms. If we think that 'scientific inquiries' can be distinguished from others by the normative standards used in carrying them out, then our question concerns whether the pragmatists thought that there were legitimate and important doubts and problems that cannot be addressed scientifically but can be addressed in some other manner. Another aspect of the issue asks whether there are issues which should be approached 'non-scientifically' even if they are susceptible to scientific treatment. We can now look at some quotations from Peirce and James which suggest that they respond to these issues in different ways.

Inquiry and the method of science

Peirce's famous paper, 'The fixation of belief' (W3: 242–57), presents a relatively unsystematic and non-technical version of his views about inquiry.[2] He asked what rules or methods we should use in trying to carry out inquiries. The most common reading of his paper suggests that he compares four different methods only one of which does not turn out to be dangerously unstable. The winner is 'the method of science.' Although this is not described in detail in this paper, Peirce does identify its 'fundamental hypothesis': we are to employ substantive rules which are answerable to this basic assumption. The 'fundamental hypothesis' runs:

> There are real things whose characters are entirely independent of our opinions about them; those realities affect our senses according to regular laws, and, though our sensations are as different as our relations to the objects, yet, by taking advantage of the laws of perception we can ascertain by reasoning how things really are; and any man, if he have sufficient experience and reason enough about it, will be led to the one true conclusion.
>
> (W3: 254)

We learn elsewhere that the method of science is a sophisticated variant of the hypothetico-deductive system accompanied by an attempt at a logic of 'discovery', and that he defends the use of induction by showing that (like fair statistical sampling) error will eventually be eliminated and truth arrived at. This is reflected in his (famous or notorious) account of truth which (in rough paraphrase) runs: If it is true that *p*, then anyone who inquired into whether *p* long enough and well enough (presumably using the method of science) would eventually arrive at a belief in *p* which would not be disturbed by further evidence or inquiry.

If the method of science is the only respectable method of inquiry, and if truth is understood by reference to a fated convergence of opinion to be found in properly conducted inquiries, it would seem to follow that any truth can be made the object of a scientific inquiry. It accords with this that Peirce thought that philosophical inquiry should be scientific. He announced his intention of bringing 'modern mathematical exactitude to philosophy' and he said that he wanted to 'rescue the good ship Philosophy for the service of Science from the lawless rovers of the sea of literature.' (*CP* 5.449, 1902) It seems reasonable to suppose that, for Peirce, the limits of scientific knowledge coincide with the limits of knowledge in general.[3]

We can note just two features of James' position which suggests that he would not agree. First, his account of truth. He offered many (not obviously equivalent) explanations of truth. But these include: 'The truth is whatever is expedient in the way of belief. Expedient in any way at all. (But expedient in the long run and for the most part, of course)' (James 1907: 106).

Many of his remarks about ethical beliefs, about belief in freedom of the will, and about belief in Gods suggest that he takes this seriously. He saw a major task of his pragmatism as to reconcile intellectual responsibility with the possibility of viewing the world with optimism (James 1907: 12–13): and he thought that neither traditional empiricism nor the contemporary Newtonian–Darwinian scientific view of the world stood much chance of doing that. We will return to intellectual responsibility later, but there are passages which suggest strongly that he thought that 'contributing to optimism' was a truth-relevant characteristic.

Secondly, we should consider the argument of his paper 'The will to believe'. Attacking W. K. Clifford's claim (Clifford 1877) that it was wrong to believe anything on the basis of 'insufficient evidence', he identified two potentially conflicting cognitive goals that we possess. We are, obviously, anxious to avoid error; and if that were our sole goal, Clifford's advice would be sound. (Although note that 'insufficient' evidence is, as its stands, rather vague). But we also desire to possess true beliefs; we need them as a basis for action: another goal is thus relief from agnosticism. The trade off between these two desiderata is clear: avoiding agnosticism increases our risk of error; and we can minimise the risk of error by remaining agnostic on many matters. James' suggests that *in the sciences*, avoidance of error is more important than relief from agnosticism, so that Clifford's demand may be acknowledged. But there are many areas of life in which possession of truth is worth the risk of error, and in such cases it may be appropriate ('on passional grounds') to believe 'in advance of the evidence'. He is clearly struck by the idea that in some cases (he refers to religion), evidence may be vouchsafed only to those who already believe the proposition that it is evidence for. And in other cases too, the urgency of the matter in which we have to act may lead us to relax or ignore evidential standards employed elsewhere in our lives..

He has two different sorts of points here. First, for practical reasons, there may not be enough time to settle a question 'scientifically': 'scientific' knowledge is *possible in principle* but, in practice we must act without obtaining it. Secondly there may be cases where insisting on scientific standards of evidence will remove the possibility of knowledge: evidence is only available to those who already believe. But notice that even in this case, he is happy to talk about 'evidence'. We may think of his idea here as related to what is often described as the theory ladenness of observation: what we see depends upon the expectations and concepts that we bring to the situation. Training is required before we can make observations using a microscope or observe and identify the tracks in a cloud chamber. Unless we know what to look for, we will not notice what is there to be seen. James carries this idea further than most. The believer will see the work of God in his observations of his surroundings; the agnostic will be blind to what is there to be seen. If you acquire a belief in God on bases other than those of evidence, this will enable you to have the experiences that will confirm your belief. Another possibility is that we test the hypothesis that God exists by exploiting the idea that if he does exist, then believers will have fulfilling lives. If he does not exist, by contrast, the believer will be disappointed and fail to find fulfilment. But, of course, if that is how we test the hypothesis, we must acquire the belief on a non-evidential basis *before* putting it to the test: acquiring the belief becomes a sort of experiment. To conclude this section, James appears to agree with Peirce in identifying science by reference to the normative standards used in evaluating propositions. But he questions the general applicability of such standards.

Science

Commenting on the passage about rescuing 'the good ship philosophy for the service of science from the lawless rovers of the sea of literature' (CP 5.449), Susan Haack has correctly noted that 'it should not be taken as suggesting that philosophy is parasitic on, or that it should be replaced by, the natural sciences' (Haack 1996). The point is, rather, that philosophy should become scientific. And this means not only that it should use the method of science, but also, more importantly, that it should be undertaken with the 'the scientific attitude'. Both 'method of science' and 'scientific attitude' have a wider application than merely to work in the special sciences. Haack continues: 'Part of the point of Peirce's insistence that philosophy should become scientific is, precisely, that there is an attitude of mind and a method of inquiry manifested, not invariably or exclusively, but primarily, by natural scientists, which all inquirers can and should adopt.' This scientific attitude is described in such terms as: 'the passion to learn' and 'an intense desire to find things out'.

It is further explained: 'Science does not so much consist in *knowing*. not even in organised knowledge, as it is in diligent inquiry into truth for truth's sake, without any axe to grind . . . from an impulse to penetrate into the reason of things' (CP 7.49).

The scientific attitude then, involves a spirit of disinterested truth seeking. And the scientific method involves reasoning and testing hypotheses against experience: proposing hypotheses, making careful observations , carrying out 'experiments' on the basis of our understanding of the hypotheses we wish to test, and then noting the results of those 'experiments'.

According to Peirce in 'The fixation of belief' we all use this 'method' unthinkingly much of the time anyway, although few of us adopt the 'scientific attitude'. So the 'method of science' is not something that contrasts with everyday and commonsense ways of solving problems. Serious scientific work involves reflecting upon the methods employed and using them 'diligently' and responsibly, exercising rational self control in pursuit of the goal of the disinterested pursuit of truth. Observations are made deliberately and on the basis of reflection rather than habitually and unthinkingly. But such reflection is a possibility outside the special sciences and Haack appears to read Peirce as recommending that we adopt this as a general policy: rational self controlled inquiry is inquiry using the method of science and undertaken in the scientific spirit.

This supports the view that, for Peirce, science has no limits: any problem that can be solved can, and should, be approached scientifically. But – and it is worth emphasising this – this is compatible with a rejection of the sort of 'scientism' which is often associated with pragmatism. Although Peirce wants his philosophy to be scientific, he also thinks that the natural and social sciences have very little (possibly nothing) to contribute to logic and philosophy.[4] Why he should think this is something we will turn to in a moment. The current question concerns *how* he can think this; what remains for a scientific philosophy to be? A crucial difference between the special sciences and philosophy concerns the *kinds of observations* that they employ. Experimental scientists rely upon special observations, ones that exploit experimental techniques and instruments which themselves reflect our advanced stage of scientific knowledge. Cloud chambers will serve as an example. Philosophy, by contrast, employs no special techniques of experiment and observation. Instead, it relies upon 'those observations which every person can make in every hour of his waking life'. He describes philosophy as that 'branch of positive science' which 'does not busy itself with getting facts, but merely with learning what can be learned from that experience which presses in upon every one of us hourly and daily' (CP 5.120, 1903). The main obstacle to philosophical progress is the fact that most of these 'facts' are *so* obvious, *so* familiar, that they easily escape our notice. Parallels with Wittgenstein and with some phenomenological techniques should be clear.

The knowledge we are supposed to obtain through philosophy stands in a rather ambivalent relation to questions about scope of science. Philosophical inquiry employs the method of science, and it is carried out in the scientific spirit, but it does not seek the sort of knowledge that could be provided by one of the special sciences. Indeed, results from the special sciences will always be too tentative, too fallible to answer to our more philosophical needs. If philosophy is to ratify the normative standards that are employed within the special sciences, it will be powerless if its results have the tentative and revisable character which is typical of results at the scientific cutting edge.

Let us collect our thoughts and note three progressively stronger (more restrictive) conceptions of an inquiry being pursued in a scientific manner:

1 The inquiry is carried out through 'reasoning and experience'; the beliefs arrived at are determined by what we make of the experiential information we collect and how we reason with it.
2 The inquiry is also carried out in the 'scientific spirit'; we seek to reason responsibly and diligently and we are motivated (primarily or entirely) by a desire to reach an opinion which is true.
3 The inquiry relies upon 'special observations' and the techniques of experimentation, measurement and evaluation associated with the special sciences.

According to Peirce, philosophy meets the first two of these conditions. His position differs from the views associated with contemporary philosophers who work in 'naturalised epistemology' because he denies that it should try to meet the third condition.

William James on science and responsible inquiry

I now want to note some passages from James which suggest that he too thinks that responsible inquiries should be guided by experience and reasoning. Three points are immediately relevant. First let us consider some more of his formulations of his claims about truth. He claims that a true proposition is one that 'leads us prosperously from one part of our experience to another': we desire beliefs that will 'put us into a satisfactory relation to our experience.' Moreover when considering how we deal with doubts, resolving tensions and problems created by the arrival of new experience, he offers some epistemological proposals that resemble Quine's insistence (Quine 1974: 137) that the fundamental standards of epistemic appraisal are conservatism and empiricism: we are to make the minimum adjustment to our corpus of beliefs which enables us (presumably in a non ad hoc manner) to accommodate the surprising experience.

We should note too the philosophical doctrine that James announced in the preface to his book *The Meaning of Truth*. This view – 'radical empiricism' – became the foundation of his later metaphysical system. It begins with a 'postulate': 'The only things that shall be debatable among philosophers shall be things definable in terms drawn from experience (Things of an unexperiencable nature may exist *ad libitum*, but they form no part of philosophical debate)' (James 1909: 6–7).

The other elements of the position involve an insistence that experience is a good deal *richer* than traditional empiricists have assumed: he claimed that parts of experience are held together by relations that are themselves parts of experience (a view to which Peirce was sympathetic although he would express it by saying that experience was continuous and that causal and nomological relations between parts of experience were themselves reflected in the continuity of experience).

A third observation. James explicitly turns to a sketchy view about the character of scientific theory in order to defend his view of truth. He presents his theory as a general application of the notion of truth which is applicable in the sciences. But here there is a marked difference with Peirce. Where Peirce appears to be a scientific realist (he describes himself as a realist 'of a somewhat extreme stripe' (*CP* 5.470) and the fundamental hypothesis of the method of science strongly suggests that this is his view) James's view of science is more instrumentalist.

> Scientific logicians are saying on every hand that these entities and their determinations, however definitely conceived, should not be held for literally real. It is *as if* they existed; but in reality they are like coordinates of logarithms, only artificial short cuts for taking us from one

> part to another of experience's flux. We can cipher fruitfully with them; they serve us wonderfully; but we must not be their dupes.
>
> (James 1909: 92)

Comparing the different versions of the world provided by science and by common sense, he concludes:

> There is no ringing conclusion possible when we compare these types of thinking, with a view to telling which is the more absolutely true. Their naturalness, their intellectual economy, their fruitfulness for practice, all start up as distinct tests of their veracity, and as a result we get confused. Common sense is *better* for one sphere of life, science for another, philosophical criticism for a third; but whether either is true absolutely, Heaven only knows… They are all ways of talking on our part, to be compared solely from the point of view of their *use*. The only literally true thing is *reality* (sic); and the only reality we know is . . . sensible reality, the flux of our sensations and emotions as they pass.
>
> (James 1907: 93)

In the background here is a functionalist view of concepts and ideas: an idea (functionally considered) is an instrument for enabling us the better to have to do with its object and act upon it. (It may be an irony that the functionalism about concepts and ideas employed here is probably shaped by the functionalist theory of mind developed in James' major contribution to natural science, his *Psychology*. In fact James' theory of truth is a fairly direct corollary of this account of thoughts and ideas: if an hypothesis is an instrument, then it is true if it achieves its purposes well, if it works well.

If all ideas and hypotheses are indeed instruments for coping with experience, and in this respect non-scientific ideas do not differ from scientific ones, James need not object to the view that all problems can be solved scientifically in the first of our three senses. But the idea that 'science' may not meet all our needs (James considers the suggestion that invaluable common sense categories such as colours may have no place in the scientific view of the world) opens up space for Rorty's appropriation of James's name for the view that the scientific version of things is just one among others.

What then should we take away from these comparisons? First, Peirce's philosophy of science has a much more *realist* flavour than James'. The latter views theories (like other beliefs and concepts) as cognitive instruments, as tools for navigating our way around our experience, making effective predictions and rendering our surroundings intelligible. The question whether 'theoretical entities' such as protons and strings of DNA 'really' exist would be dismissed as a pseudo-question. For Peirce, by contrast, the sciences reveal invisible mechanisms that are operative in the production of observable phenomena. Second, Peirce's emphasis on the scientific *attitude* – on the need for diligent, responsible, self controlled inquiry – and (most important) on *disinterested inquiry* establishes one contrast between scientific and non-scientific inquiry which is not echoed in the quotations from James that I have offered here. However we should recall James' insistence that preferring agnosticism to error is one mark of scientific minds which is not shared with non-scientific styles of inquiry.

It is connected to this that the notion of truth as convergence has a fundamental role in Peirce's picture: our goal is to arrive at the opinion which sufficient responsible inquiry is *fated* to reach. Once this is our target, we can see how the other features of the scientific spirit are required. This seems to be what the scientific attitude requires as a target. Interestingly, James does articulate a notion of 'absolute truth' which is almost identical to Peirce's. But he denies that this has a normative role in the ways in which we plan inquiries and evaluate opinions. Granted his adherence to an instrumentalist view of science, this may not be surprising. We have now raised two issues about the proper scope of scientific knowledge. The first concerns the proper signification of 'science': we have seen that, for Peirce and probably for other pragmatists, this can have a wider or narrower meaning. We can allow that all beliefs should be grounded in experience while denying that inquiries carried out using the scientific method or guided by the scientific spirit are always in order; and we can urge that philosophy be scientific without expecting it to make use of information from the special sciences. The second issue is about the relations between 'scientific' knowledge and the systems of beliefs and categories which comprise 'commonsense'. We shall now examine this second issue in more detail. How do science and commonsense differ?

Science/commonsense/practice

In The *Will to Believe*, after linking Clifford's strategy to what goes on in the institution of science, James complains:

> He who says 'better go without belief forever than believe a lie' merely shows his own preponderant horror of being duped.
>
> . . . Science has organised this nervousness into a regular technique – her so called method of verification; and she has fallen so deeply in love with her method that one may say she has ceased to care for truth by itself at all. It is only truth as technically verified that interests her.

> The truth of truths might come in merely affirmative form, and she would decline to touch it.
>
> . . . Moral Questions immediately present themselves as questions that cannot wait for a sensible proof... [They concern] not what sensibly exists, but of what is good, or would be good if it did exist. Science can tell us what exists, but to compare the worths, both of what exists and of what does not exist, we must consult not science but what Pascal calls our heart. Science itself consults her heart when she lays it down that the infinite entertainment of fact and correction of false belief are the supreme good for man.
>
> (James 1897:25, 26–78, 27)

Bearing in mind James' role in putting Psychology on a scientific footing through works such as his *Principles of Psychology*, we might be struck by the crudity of some of these remarks about science. His own work is scientifically serious but betrays no neurotic fear of error or obsession with 'infinite ascertainment'.

When contrasting our other cognitive needs with those served by the sciences, he produces examples about *values*. It is not clear whether he holds that a scientific answer might be available in principle to these questions, but that their urgency prevents our tackling them through detached scientific inquiry. More likely, it is his view that a fully scientific solution to these problems is simply not possible. Secondly we should note that he sees a role for sentiments, for appeals to the heart, in dealing with these evaluative issues.

It is interesting that Peirce makes some very similar remarks. They are found in texts from all stages of his career, but receive their clearest (and possibly most exaggerated) statement in some lectures delivered in Cambridge, Massachusetts shortly after James' famous lecture on the will to believe was delivered. On this occasion he is concerned with how we should go about answering 'vital questions'. Invited to lecture on philosophical responses to 'vitally important topics' he insisted that the search for answers to such questions was a search for belief: 'full belief' (*CP* 1.635).[5] But this kind of belief – 'what is properly and usually called belief' – 'has no place in science at all'. This is because the scientific spirit requires us to be ready 'to abandon one or all' of our results 'as soon as experience opposes them'. If beliefs are implicated in our responses to issues of vital importance, we shall not possess this readiness to abandon them. In such passages, he agrees with James that science should prefer agnosticism to error, looking upon favoured theories simply as what it is rational to assert at the current stage of our progress towards the truth. Although the logic of science can explain why the scientific method should take us to the truth *eventually*, it provides no grounds for assurance that it can help us to reach the truth soon. For someone to pursue science for the sake of reaching believable results which will have practical applications 'would spoil him as a scientific man'. (*CP* 1.619) Our practical 'interest' in our results would interfere with the readiness to drop previously attractive theories which the scientific spirit requires. It is a 'treason against reason' to turn to theory, whether drawn from science or philosophy, for help with the important questions of life.

How then should we tackle 'vital questions', issues of ethics, practical politics, religious belief or choice of life project? In the lecture we are discussing he announced that 'in everyday business, reasoning is tolerably successful, but I am inclined to think that it is done as well without the aid of theory as with it' (*CP* 1.623):

> Men many times fancy that they act from reason when, in point of fact, the reasons they attribute to themselves are nothing but excuses which unconscious instinct invents to satisfy the teasing 'why's of the ego. The extent of this self-delusion is such as to render philosophical rationalism a farce.
>
> (*CP* 1.631)

Matters of vital importance are to be settled with the aid of 'instinct' and sentiment rather than theory (Hookway 2000 chapters 9 and 10). This theme had been present from Peirce's earliest writings: from the early 1870s he scorned those who expected theologians and metaphysicians to contribute to our understanding of religious matters, and his late 'proof' of God's reality depends crucially on the claim that religious belief comes 'naturally' to those who are open to it.

As with James, it is difficult to see how radical these claims are in emphasising the limits of the trust we should place in theory and the spirit of science. Is he making the practical point that many of these vital matters are urgent, calling for answers far more quickly than science can guarantee to deliver them? Or is he concerned with the more radical claim that, in the face of these sorts of issues, inquiry in the scientific spirit would never provide what we want? The question is difficult to answer because of the different meanings that can be attached to an inquiry being 'scientific'. The remainder of this section will explain this.

It is initially difficult to reconcile these claims with Peirce's defence of the 'Method of science' in 'The fixation of belief'. He seems to be suggesting that when answering vital questions we should use the 'a priori method', accepting propositions which are 'agreeable to reason', a strategy which he dismissed in 1877 as potentially very unstable, as exposing belief to the vagaries of fashion. This appearance can be resisted, however. As we have seen above, three related, but distinct elements enter Peirce's view of the scientific method. First, beliefs are to be grounded in experience, our opinions being determined by the effects of the things on our senses. Second, serious scientific inquiry should be carried out in 'the scientific spirit': we seek to reason responsibly and diligently and we are motivated (primarily or entirely) by a desire to reach an opinion which is actually true. Finally a distinction was drawn between the experiential resources to be used in different kinds of inquiries: the special sciences rests upon 'special observations' using refined techniques for measurement and for constructing experiments. Philosophical inquiries may meet the first two conditions but not the third: we employ casual everyday observations of what is available to everyone. The difficulty of many philosophical problems stems from the fact that such phenomena can be so familiar, so ubiquitous a part of our lives, that we fail to take explicit notice of them. I suspect that Peirce would restrict his claim that belief has no place in science to the special sciences.

Our responses to 'vital questions' can then be 'scientific' in two different respects. First the dominant philosophical influence at Harvard when Peirce and James were students was the Scottish philosophy of commonsense. As this doctrine was incorporated into pragmatism, it emphasised a body of 'instinctive' vague common sense certainties which reflected an indefinite mass of experience spread over many generations. Forming part of our general scientific or non-scientific culture, inculcated through training and education, manifested in 'sentimental' or emotional judgments of irrationality or plausibility, they were recognised as a powerful yet fallible tool for prediction and understanding. As well as guiding us in dealing with vital questions, they could have an important role within science, especially in the logic of discovery, the theory of 'abduction'. It is rational to adopt an attitude of qualified trust in our intuitive judgements of the plausibility of theories and conjectures. Is the knowledge expressed in such common-sense certainties 'scientific'? It is built on experience and does not reflect the results of putative faculty of (for example) intellectual intuition. But if we try to inquire into the matters with which it deals using 'scientific' reflection, we risk losing the knowledge that it embodies. Insisting that we only use our collective 'experience' when we can turn it into a precisely formulated body of theory that can be tested rigorously risks abandoning the accumulated wisdom of our traditions, including the scientific tradition.

But, secondly, our reliance upon these commonsense certainties is not uncritical or unreflective. Indeed Peirce designated his own mature philosophy 'Critical common-sensism' (*CP* 5.438–452).[6] Philosophical reflection itself can help us to understand the wisdom of a cautious dependence upon common-sense and instinctive belief. Empirically grounded inquiry can reassure us that this is the best way to deal with our vital problems. Like James, Peirce wanted no knowledge that was not based upon experience. But, also like James, he was aware that insisting that all inquiries should be driven by the scientific spirit, and that all problems could, in principle, be solved by inquiries of this kind, threatened to lose knowledge and leave us with an inadequate guide to life. Indeed it would also leave us without many of the norms and attitudes that are required for scientific inquiry.

Commonsense and science: Peirce and Dewey

For Peirce, science is an activity carried out using a particular method and with a particular spirit. James sometimes suggests the same. Science provides a way of approaching any issue: it is not distinguished by its subject matter. Hence philosophy can use the scientific method - it differs from 'real' science in the kinds of observations and experiments it employs.

For Dewey (and possibly James), while all responsible inquiry shares this general pattern of testing things against experience, the special sciences have a distinctive kind of *subject matter*. Commonsense problems and inquiries are concerned with 'interactions into which living creatures enter in connection with environing conditions in order to establish objects of use and enjoyment' (Dewey 1938: 115). This, we are told, has a number of important consequences. In so far as commonsense knowledge forms a system, this is 'practical rather than intellectual', and it is 'constituted by the traditions, occupations, techniques, interests, and established institutions of the group'. The 'meanings' and symbols employed will be those found in the common language of the group; they will be 'determined by the common culture of the group'. And this means that inquiries will occur within a specific cultural context: results will be ratified as applicable within that context and not as having a wider application: 'there is no such thing as disinterested intellectual concern with either physical or social matters.' Where the 'symbol-meaning systems involved are connected directly with cultural life activities and are related to each other in virtue of this connection', there is no 'science as such'.

I hope that the reader is struck by the similarities between these claims and Peirce's remarks about commonsense and vital questions. Dewey adds to Peirce's emphasis that such questions must be *answered* by reference to the concepts built into commonsense the plausible suggestion that the problems being addressed are also *formulated* in terms drawn from this 'common culture'. But science is different:

> [M]eanings are related to one another on the ground of their character as meanings, freed from direct reference to the concerns of a limited group. The intellectual abstractness is a product of this liberation, just as the 'concrete' is practically identified by directness of connection with environmental interactions. Consequently a new language, a new system of symbols related together on a new basis, comes into existence, and in this new language semantic coherence, as such, is the controlling consideration.
>
> (Dewey 1938: 115–16)

In explaining what is distinctive about scientific inquiry Dewey draws attention to two features. The first is that scientific knowledge is more general than commonsense knowledge because it is not restricted to a particular context. As he puts it, the reference of scientific theories 'is to *any* set of time and place conditions.' (Dewey 1938: 117) Secondly he claims that 'since meanings are determined on the ground of their relations as meanings to one another, relations become the objects of inquiry and qualities are assigned a secondary status' (ibid.: 116). An example will illustrate what this means. Ordinary colour concepts and other ideas of secondary qualities can be used within commonsense inquiries. Members of a 'limited group' have similar perceptual apparatus, and the context will generally be sufficiently fixed for colour ascriptions to be stable. If science abstracts from these limitations it prefers theories cast in terms of 'primary relations' such as 'position, motion and temporal span'. He concludes: 'In the structure of distinctively scientific objects these relations are indifferent to qualities' (ibid.: 116). Science dispenses with secondary qualities and is concerned with very general, abstract laws that deal with primary qualities. Our everyday lives require concepts of qualities, of things such as colours, tastes and values.

Although Dewey distinguishes commonsense and science in terms of subject matter while Peirce does so by reference to the different methods employed in solving problems, it should be apparent that this difference belies a deeper similarity of view. While solving commonsense problems, the cultural norms and standards used in formulating problems also point the way towards a solution. We are wise to trust our grasp of common meanings and allow them to shape our methods of inquiry. If we propose to step back from everyday problems to seek knowledge which abstracts from the context of these 'common meanings' and which seeks knowledge which is truly general, we need to replace the use of everyday qualitative concepts with terms for abstract relations which can be applied in a wide range of contexts. It is unsurprising that we can no longer trust our context bound habits of belief and inference, and our commonsense assumptions. Explicit reasoning and deliberation is then required. In that case, Dewey and Peirce may point to different ways of expressing a common point of view.

Conclusion

Pragmatism pulls us in two directions. First, it is marked by a strong naturalistic tendency. Thus Dewey emphasises the similarities between the patterns of inquiry found in sophisticated science and in simpler forms of animal cognition and he seeks a philosophy that learns the lessons to be drawn from the Darwinian revolution. William James' contributions to psychology led to a functionalist account of cognition and an instrumentalist view of concepts that fuelled the distinctive form taken by his own pragmatism. And his 'radical empiricism' introduced a theme shared with Peirce: philosophy must be grounded in experience rather than in *a priori* speculation; but, as psychology teaches, experience is far richer than earlier atomistic forms of empiricism had allowed. When Peirce came to see that his pragmatism required a distinctive metaphysical view, one which endorsed an anti-Humean form of realism about law, generality and causal interaction, he sought a system of 'scientific metaphysics'. However, this did not mean that metaphysics and cosmology were to be derived from current results in special sciences such as physics and biology. They were to be based on experience and on controlled 'scientific' methods of investigation.

Second, it was recognised that the special sciences were marked by especially rigorous standards of evidence, by distinctive attitudes towards current results and by a search for results that were not tied to specific contexts of investigation. This means that in connection with many of the most important concerns of life, investigations that ape the methods of the special sciences have little relevance and can present substantial obstacles to understanding and fulfilment. Peirce emphasised the wisdom which is present in commonsense and in instinctive habits of feeling and inference, wisdom that is lost if we trust conscious reflection and deliberation in trying to settle important ethical dilemmas or matters of religious belief. James added to this celebration of commonsense a stress upon the variety of interests and purposes that our concepts and beliefs can serve, upon the need for beliefs that will meet our needs in the particular contexts in which we find ourselves, and upon the rationality, in some circumstances, of belief 'in advance of the evidence'. And Dewey claimed that inquiries typically respond to indeterminacies in our current situations, coming to an end when something analogous to aesthetic 'judgment' confirms that our situation is whole and newly unified. Scientific rigour, and its concern with abstract systems of relations rather than familiar everyday 'qualities', may often be an obstacle to restoring the wholeness of our cognitive 'situation'.

Notes

1 These differences are discussed in more detail in Hookway 1997.
2 The apparent clarity and ease of this paper is a surface phenomenon. Some of the complexities it involves are discussed in 'Belief, confidence and the method of science', chapter 1 of Hookway 2000. A useful and accessible survey of interpretative issues is found in Anderson 1995: 82–117. One source of the difficulty, and of the apparent ease, is that the concern with philosophical architectonic which characterises most of Peirce's work – including his theory of categories and his systematic theory of signs – are largely absent.
3 Issues raised by this theory of truth and convergence are discussed at length in chapters 1 to 4 of Hookway 2000.
4 From his earliest writings and lectures, Peirce attacked James Mill and John Stuart Mill for thinking that *psychology* was relevant to work in logic and epistemology.
5 A full text of these lectures is in *Reasoning and the Logic of Things* (1992). The passages cited here are all from the first lecture. They are discussed in Hookway 2000 chapter 1.
6 I have discussed Peirce's critical common-sensism more fully in Hookway 1985: 229 ff. and 2000 chapter 8.

Bibliography

Anderson, D. (1995) *Strands of System: The Philosophy of Charles Peirce*, West Lafayette: Purdue University Press.

Clifford, W. K. (1877) 'The ethics of belief', *Contemporary Review* 29: 283–309

Dewey, J. (1938) *Logic: The Theory of Inquiry*, Boston: Holt, Rinehart and Winston.

Elgin, C. (1997) *Between the Absolute and the Arbitrary*, Ithaca: Cornell University Press.

Haack, S. (1996) 'As for that phrase "Studying in a literary spirit"', *Proceedings and Addresses of the American Philosophical Association*, vol. 69.

Hookway, C. J. (1985) *Peirce (The Arguments of the Philosophers)*, London: Routledge.

Hookway, C. J. (1997) 'Logical principles and philosophical attitudes: Peirce's response to James's pragmatism', in R. A. Putnam (ed.) *The Cambridge Companion to William James*, Cambridge: Cambridge University Press:145–65.

Hookway, C. J. (2000) *Truth, Rationality and Pragmatism: Themes from Peirce*, Oxford: Oxford University Press.

James, W. (1897) *The Will to Believe and Other Essays in Popular Philosophy*, reprinted Cambridge: Harvard University Press, 1979.

James, W. (1907) *Pragmatism: A New Name for Some Old Ways of Thinking*, reprinted Cambridge: Harvard University Press, 1975.

James, W. (1909) *The Meaning of Truth: A Sequel to 'Pragmatism'*, reprinted Cambridge: Harvard University Press, 1975.

Mounce, H.O. (1997) *The Two Pragmatisms*, London: Routledge.

Peirce, C. S. (1931–58) *Collected Papers of Charles Sanders Peirce*, ed. C. Hartshorne and P. Weiss (vols 1–6) and Arthur Burks (vols 7–8). Referred to as '*CP*' followed by volume and paragraph number.

Peirce, C. S. (1982) *Writings of Charles S. Peirce: A Chronological Edition*, ed. M. Fisch *et al.*, Bloomington: Indiana University Press. Referred to as 'W' followed by volume and page.

Peirce, C. S. (1992) R*easoning and the Logic of Things*, ed. K. L. Ketner, Cambridge, Mass.: Harvard University Press.

Quine, W. V. D. (1974) *The Roots of Reference*, La Salle, III.: Open Court.

Rorty, R. (1982) *Consequences of Pragmatism*, Hassocks: Harvester.

8 Husserl and the crisis of the European sciences

Dermot Moran

In this essay I want to sketch Husserl's general philosophical concerns, focusing in particular on his contribution to the critical evaluation of the scientific enterprise. There is the widespread belief that the philosophical traditions in Continental Europe in the twentieth century have been broadly anti-scientific in orientation, and, therefore, it is assumed that Edmund Husserl (1859–1938), since he influenced such European philosophers as Gadamer, Heidegger, Marcuse, Horkheimer, Merleau-Ponty, Sartre, Levinas, Ricoeur, and Derrida, among others, must be an historical source of this anti-scientism.[1] On the other hand, Husserl was initially primarily known for his *Logical Investigations* (1900–1), which, in Germany, provided the deathblow to then current psychologistic interpretations of logic, Frege's own efforts in the same area being in obscurity at that time.[2] Students of the history of analytic philosophy recognise Husserl's refutation of logical psychologism and his strong defence of the ideal objectivity of propositional contents as on a par with the contribution of Frege in the clarification of the essential nature of logic. Moreover, Husserl belonged to the new wave of logicians at the end of the nineteenth century who fully acknowledged that logic was actually a part of mathematics.[3] Thus Richard Rorty correctly links Husserl and Russell together as two paradigmatic figures seeking to recapture the mathematical spirit in philosophy (Rorty 1980: 166).[4] But even those who acknowledge Husserl's historical contribution to modem logic hold that, in his later works, he failed to take advantage of the mathematical formalisation of logic in order to analyse the nature of language and thought, and, indeed, strongly opposed the growing technicisation (*Technisierung*) of the discipline of logic. Furthermore, Husserl's antipathy to psychologism and to reductive naturalism led him to question the impact of modern mathematical sciences on the human cultural world, so that, in his later work at least, he can be seen as inviting and encouraging the anti-scientism which has come to characterise recent Continental philosophy in general.

It is certainly true that, in his later writings, especially the *Crisis of European Sciences* (1936), Husserl proposed a radical rethinking of the manner scientific practice was distorting our normal engagement with the familiar 'life-world' (*Lebenswelt*). His aim was not to reject science in favour of the '*Schwärmerei*' of some kind of irrationalist existentialism, but rather to make the sciences more fully and self-consciously rational, to separate pure science from an ideologically distorted scientism. Husserl always opposed irrationalist *Lebensphilosophie* and, in opposition to the existential deformation of phenomenology proposed by Martin Heidegger, declared: 'For me, philosophy, as an idea, means universal, and in a radical sense, "rigorous", science' (Husserl 1989: 406, Hua V 139). Indeed, Husserl was driven to breaking his ties with his favoured successor, Martin Heidegger, after the publication of *Being and Time* (1927), precisely because the latter had misunderstood the scientific nature of Husserl's project and had reduced it to 'anthropology'.

Husserl's entire working life was a struggle to make sense of what *science* (*Wissenschaft*) means, both as a *practice* and as an *ideal*. For Husserl, science is an open field of 'infinite tasks' standing before us (Husserl 1970a: 279, Hua VI 324). But one must carefully distinguish between the motivating and guiding *ideal* of science and its various actual historical forms of practice. Even when he enigmatically declared in 1935 that for philosophy as science 'as serious, rigorous, indeed apodictically rigorous science – *the dream is over* (*der Traum ist ausgeträumt*),' this must be understood not as Husserl's abandonment of his ideal, but rather, as his recognition that the ideal itself now needed to be understood and located in history and its motivating force clarified, since it had been effectively abandoned by *Existenz* philosophers such as Jaspers and Heidegger (Husserl 1970a: 389, Hua VI 508). Husserl himself would have been shocked to be considered anti-scientific in his philosophical outlook, since, for him, science was the embodiment of rational practice and, as such, the only hope for the salvation of humanity.

Husserl's scientific formation

Husserl's own background was steeped in the sciences. While still a Gymnasium student at Olmütz and Vienna, Husserl displayed an aptitude for mathematics and expressed an interest in studying astronomy. At the University of Leipzig, he studied mathematics, physics and astronomy; before moving to Berlin, where he studied mathematics with some of the leading mathematicians of the day, including Leopold Kronecker (1823–1891) and Karl Weierstrass (1815–1897). It was Weierstrass who introduced Husserl to Bolzano's *Paradoxes of the Infinite*. Throughout his life, Husserl constantly cited approvingly the formative influence of Weierstrass, and even remarked that he had intended to do for philosophy what Weierstrass had done for mathematics. In 1882 in Vienna, Husserl submitted his doctoral thesis on the calculus of variations, a branch of differential calculus, supervised by Leo Königsberger, another student of Weierstrass.

Soon after completing this doctorate, Husserl was converted from a career in mathematics to philosophy by the charismatic Franz Brentano, whose lectures he attended in Vienna from 1884 to 1886. Brentano's ideal of exact philosophy, philosophy as a rigorous science, strongly appealed to the young mathematician, who himself had become interested in foundational questions concerning mathematics and logic. Originally, Husserl was particularly drawn to Brentano's programme for the reform of Aristotelian syllogistic logic, but he was soon completely won over to Brentano's anti-speculative conception of philosophy including a strong preference for Hume and Berkeley over and against the 'mysticism' of the German Idealists (a preference which is still present in Husserl's *Crisis*). Brentano also greatly admired the positivists, especially Auguste Comte, and Husserl too closely followed the work of Ernst Mach and Avinarius, as well as Brentano's students Carl Stumpf, Ehrenfels and Anton Marty.[4] Thus, the philosophical atmosphere Husserl encountered in Austria was strongly realist, with huge respect for the achievement of the positive sciences. The *Ernst Mach Verein*, and later the Vienna circle, also grew from the same philosophical sources. Indeed, the Manifesto of the Vienna Circle, in its section on historical influences, cites Brentano approvingly for his project of the reform of logic and his opposition to Neo-Kantian idealism. Having spent two years with Brentano, Husserl then moved to Halle to habilitate with Carl Stumpf who was intent on developing Brentano's descriptive psychology through careful analysis of sense perception and spatial awareness. With Stumpf, Husserl wrote his Habilitation thesis, *On the Concept of Number, Psychological Analyses*. Georg Cantor, the founder of set theory, was one of the examiners for this thesis.

During his *Privatdozent* years at Halle (1887 to 1901), while continuing to investigate the foundations of mathematics and logic, Husserl became personally close to Cantor, but also corresponded with prominent mathematicians and logicians, including Gottlob Frege.[5] Indeed, Husserl was one of the few German philosophers at that time to acknowledge and critically discuss Frege's work on arithmetic. Later, from 1901 to 1916, as professor at Göttingen, a renowned centre of mathematics at that time, Husserl belonged to a group of distinguished researchers which included David Hilbert (1862–1943), Hilbert's assistant Richard Courant (1888–1972) and Felix Klein (1849–1925), and shared with them the view of the goal of nomological science as universal formal axiomatisation.[6] The historian of science, Alexandre Koyré, also studied with Husserl at that time, and may have later influenced Husserl's conception of modern science as a Galilean enterprise. Husserl made the acquaintance of many of the important scientists of his day, including the mathematician Brouwer and the physicist Hermann Weyl, and his Freiburg seminars drew such visitors as Rudolf Carnap (in 1924–5) and the logician William Kneale. Even though Husserl was preoccupied with developing his new science of phenomenology, throughout his entire career he continued to think critically about the nature of mathematics and geometry, as well as the origin of our spatial and temporal concepts. In his post-retirement years (1928–38), he revisited his original problematic concerning the nature of logic (e.g., *Formal and Transcendental Logic*, 1929), and, in the thirties, began a new exploration of the manner in which modern science, in carrying out the scientific programme first proposed by Galileo, idealised and objectified the world of nature and contributed to the divorce between science and the human life-world. There is a sense, then, in which Husserl's entire life's work can be seen as a meditation on science, and specifically on the nature of logic as providing the framework for science as such. Indeed, Husserl, following Bolzano, initially took logic to express the very essence of science, though he later came to see philosophy as fulfilling that role.

The ideal of scientific knowledge

For Husserl, the philosophical life, the life of contemplation, *theoria*, is the scientific life *par excellence*. Philosophy, therefore, is not only deeply and self-consciously scientific, but it is also the science which 'satisfies the loftiest theoretical needs and renders possible from an ethico-religious point of view a life regulated by purely rational norms' (Husserl 1965: 71, Hua XXV 30). Philosophy is 'humanity struggling to understand itself' (Husserl 1970a: § 5 p. 14, Hua VI 12), and science is the 'self-objectivation of human reason' (Husserl 1969 Intro p. 5, Hua XVII 4). Philosophy aims to 'elevate mankind through universal scientific reason' (Husserl 1970a: 283, Hua VI 329). Husserl's philosophical orientation was rationalist in the deepest sense of the term, aiming to achieve 'a life of universal self-responsibility'(Husserl 1970a: 338, Hua VI 272). Thus, at the end of his life, he claimed to speak as one 'who had lived in all seriousness the fate of a philosophical existence' (Husserl 1970a § 7 p. 18, Hua VI 17).

Husserl, following in the German tradition, understands by 'science' (*Wissenschaft* or, in Greek, *episteme*) all intellectually-grasped, organised knowledge, and, all through his life, he held an ideal of science as a systematically, internally related series of propositions expressed with clarity and exactness. Furthermore, science as such includes both the natural and the cultural sciences. Husserl was well aware of the disputes going on in German thought (in Rickert, Windelband and Dilthey) regarding the status of the human sciences versus the natural sciences, but, for Husserl, this particular dispute was misconceived and symptomatic of a deeper issue which had to be resolved, namely, the clarification of the ideal of science in general and the fight against a *naturalistic* misinterpretation of the natural sciences which has also distorted the understanding of the social sciences. Husserl not only wanted to avoid the split between the natural and the social sciences, he also wanted to overcome the even more pernicious gap between everyday life and the increasingly powerful knowledge of the specialised sciences. The distinction between natural and cultural sciences was itself a product of a certain institutionalisation of a schism between subjectivity and objectivity which Husserl wished to challenge through his new science of phenomenology. The central aim of Husserl's new science of phenomenology was to allow theoreticians of all kinds to see the true nature of the insights of their disciplines in an unprejudiced manner and thus phenomenology would provide a ground (*Boden*) for modem objective science. For the early Husserl of the *Logical Investigations*, phenomenology offers a critique of science through the clarification of its essential concepts. The aim is to make science more consistently and transparently scientific, removing all reliance on confused and unclear concepts. Later on, as phenomenology came to be equated with the whole of philosophy, Husserl's phenomenology sought, in the words of Merleau-Ponty, to measure the distance between human experience and science (Merleau-Ponty 1964: 29).

Husserl's concept of science as an *ideal* was an amalgam of Platonic and Bolzanian conceptions. Philosophy, for Husserl, is in essence 'theory of science' (*Wissenschaftslehre*), a term taken from Bolzano, used to express the view that philosophy is a systematic reflection on the meaning of science as such (Husserl 1969: 13, Hua XVII 127). Husserl was not alone in adopting Bolzano's conception of *Wissenschaftslehre*; indeed this term appears in Carnap's earlier works, before being replaced by the term 'logic of science' (*Wissenschaftslogik*). Husserl's conception of 'genuine science' (*echte Wissenschaft*) or 'full science' (*volle Wissenschaft*), and his understanding of the relation between philosophy and the sciences, offers a powerful alternative to the naturalistic view (expressed by Quine and his followers) of philosophy as continuous with science (Husserl 1970a: § 34 p. 124, Hua VI 127). Husserl was critical of prevailing naturalistic programmes found in *positivism* (Comte, Mach, Avinarius). But his antinaturalism did not push him in the direction of historical *relativism*. He was implacably hostile not only to emotional, irrationalist life philosophy, as we have seen, but also to various forms of historicism which led, in his view, to cultural relativism, to the collapse of the scientific ideal, and hence – in his view – to barbarism.

For Husserl, the ideal of genuine science – first formulated in philosophy but still requiring to be exemplified by philosophy – must provide the guiding idea (*telos*, *Zweckidee*) for all knowledge which seeks universality (Husserl 1973; Hua XIII 214, 217). For Husserl, science is universal, and as such intersubjectively graspable; what is true must in principle be knowable 'by anyone' (*für Jedermann*) (Husserl 1973: 293, Hua VI 329). There is no room in any science (and especially not in philosophy which gives birth to and sustains the very idea of a science) for private opinions, for individual standpoints and constructions (Husserl 1965: 74, Hua XXV 5). Science is a *Lehrsystem*, a system of teachable truths (Husserl 1970b: 250, Hua XIX/I A5).[7] Science is stored up in a system of interconnected theoretical propositions or statements, and of course, in the normal run of events, scientists merely manipulate these theoretical truths without insight (Husserl 1983: § 66 p. 152, Hua III/I 124). Thus, in calculation one is able to manipulate symbols without giving thought to what they stand for. The key to genuine scientific knowledge, however, is that all the essential insights or rational commitments which gave birth to the knowledge must in principle be reiterable (Husserl 1970a: 304, Hua VI 281). Genuine science must be able to trace back any set of claims to the original acts of evident cognition which engendered them in the first place. This is one of the Cartesian elements in Husserl. But not only must it be possible to recover the founding insights of a discipline; these insights must be intersubjectively communicable and shareable. Science lives on only in the community of practitioners of science.

In talking about scientific knowledge and cognition (*Erkenntnis*), there is always an ambiguity between the 'body of knowledge', that is, the *set of theoretical propositions* which set out the scientific knowledge, and the *cognitive acts* that grasp this knowledge. While Husserl, with his critique of psychologism and later of naturalism, is emphatic on the need to distinguish carefully between these two domains, he also wants to account for their interconnection. A complete science must have an account of how acts of cognition *grasp* their theoretical objects, and this science is what Husserl calls 'phenomenology'. Husserl's aim in fact sounds somewhat paradoxical to ears which associate pure objectivity only with natural science: Husserl wished to achieve a genuine science of subjectivity, a fully grounded objective science of the subjective, one which recognised the meaning-constituting role of subjectivity as well as the objectivity of constituted meanings in themselves. In other words, Husserl's point, against Daniel Dennett and other 'heterophenomenologists' who deny the explanatory value of first-person experience (what Husserl calls the 'subjective relative'), is that true objectivity is found not by excluding subjectivity but precisely by taking it into account. Husserl wanted scientific knowledge grounded and clarified by a fundamental science of the act of cognition itself, without lapsing into psychologism: that is, without reducing cognition to a natural process in the world.

Though philosophy had contributed to Western culture both the ideal of objectivity and the related conception of knowledge as science, Husserl had to confront two significant problems: philosophy has failed to become a science; and the special sciences have cut themselves adrift from the philosophical ideal of universal knowledge.[8] Husserl shared with Kant the conviction that philosophy to date had failed to live up to its own self-declared aim of becoming a science and remained mired in competing systems, disputation and confusion, which could lead only to moral degeneration and cultural collapse (Husserl 1970a: § 7 p. 17, Hua VI 15). In the *Crisis* Husserl again notes the repeated failures of metaphysics versus the steady march of positive sciences (Husserl 1970a: § 4 p. 11, Hua VI 8–9). Husserl's response is also articulated in Kantian terms: the achievement of a *critique of reason* (Husserl 1994: 493, Hua XXIV 445).[9] The foundations of reason (e.g., the fundamental nature of logic and ethics) remained unclarified and hence left room for endless misinterpretation. Husserl was convinced that partisan philosophy was at an end and that there was no going back to selfish business of constructing philosophies as world-views. Hume and Kant sought the reasons for the failure of metaphysics. For Husserl, Hume was the first transcendental philosopher (Descartes discovered the transcendental domain but did not enter into it) because he more than anyone else recognised how naively we assume objectivity (e.g. of the causal connection) without considering the manner it is constructed by us (Husserl 1970a: § 26 p. 96, Hua VI 99). For Husserl, it is Hume more than Kant who shows causation to be belong to the form of subjective experience rather than to the objective world.

The emergence of the scientific ideal

Husserl had a strong appreciation of the specifically theoretical orientation of pure science, remote from practical or applied interests. Indeed Husserl recognised that the '*theoretical praxis*' of philosophy and the sciences emerges quite late in the historical evolution of humanity and its nature is as yet little understood (Husserl 1970a: § 28 p. 111, Hua VI 113).[10] For Husserl, the guiding ideals of scientific *praxis*, i.e. truth in itself, pure objectivity, have been distilled from the philosophical tradition. These ideals of objectivity and truth are essential to conceiving a new and higher level of human life, universal life guided by rational ideas (Husserl 1970a: 336, Hua VI 270).[11] It was the ancient Greeks who first proposed a 'humanity which seeks to exist, and is only possible, through philosophical reason' (Husserl 1970a: § 6 p. 15, Hua VI 13). For Husserl, the Greek Enlightenment presents the first breakthrough into what is essential to humanity as such. Philosophy, then, is not an inevitable outpouring of the human spirit, but a specific, and fragile, Western European accomplishment. The Greek experience constitutes an essential part of humanity's self-awakening, and this Greek world is not merely just one form of humanity (*Menschheit*) among others, not just an empirical, anthropological type like China or India (Husserl 1970a: § 6 p. 16, Hua VI 14). The Europeanisation of other societies bears witness to this; indeed in essays written in the twenties Husserl acknowledged Japan as having joined the European scientific outlook (Husserl 1981: 326, Hua XXVII 3). According to Husserl's version of this oft-told story, the ideal of science had first been mooted by Socrates and Plato in their revolution of Greek thinking and again by Galileo, Descartes, Leibniz and others at the beginning of our era.[12] Plato, in the face of scepticism about the very possibility of attaining knowledge (Husserl refers to Gorgias's second proposition: nothing can be known), first articulated the ideal of science as a practice dominated by a purely theoretical interest which seeks to justify each step as valid in accordance with principles which have been secured in advance (Husserl 1969: 1, Hua XVII 1; 1970a: 313, Hua VI 291). Indeed, this ideal objectivity came to provide a norm for all forms of knowledge (Husserl 1970a: § 3. p. 121, Hua VI 124), but it was not until modernity that this sense of objectivity eventually produced a transformation or 'upheaval' (*Umstellung)* of our very conception of the world (*Weltbegriff*) (Husserl 1970a: 344, Hua VI 358).

For Husserl, for philosophy to come into full possession of itself, it must interrogate the manner in which this ideal of objectivity came to dominate the entire domain of human rationality, to understand the nature of its impact on our sciences, and to correct any distortions and so prevent a slide back into scepticism and irrationalism. Husserl is suspicious of Enlightenment rationality which emerged in tandem with the scientific revolution of the seventeenth century. He finds this form of rationalism naive and ungrounded and proposes its philosophical critique. But a critique of a particular form of rationality can never mean the abandonment of rationality as such (Husserl 1970a: § 6 p. 16, Hua VI 14). One must never relinquish the challenge of philosophy to be the 'possibility of universal knowledge', the vision of philosophy in the service of mankind, and philosophers as the 'civil servants of humanity' (*Funktionäre der Menschheit*) (Husserl 1970a: § 7 p. 17, Hua VI 15).

The critique of positivism, empiricism and naturalism

To achieve genuine progress through scientific knowledge, all distortions and misunderstandings of scientific reasoning must be exposed and eliminated. Just as nineteenth-century mathematics had been in crisis because it relied on different, indeed conflicting, theoretical insights to explain its various accomplishments, so too the sciences needed to be put on a secure conceptual footing. Husserl opposed two different misinterpretations of science: on the one hand, *positivism and naturalism*; and, on the other, *cultural relativism* and *irrationalist mysticism*. As we have seen Husserl admired positivism for its anti-speculative moment and for its attempt to remain true to the things themselves. Thus, in *Ideas I* (1913), he appropriated the term 'positivism' for his own phenomenological philosophy: phenomenologists are the true 'positivists' if 'positivism' means 'an absolutely unprejudiced grounding of all sciences on the 'positive', that is to say, on what can be grasped *originaliter*' (Husserl 1983: § 20 p. 39, Hua III/I 38). Nevertheless, Husserl was a severe critic of the manner in which a deficient positivism had come to dominate the scientific outlook and had rigidified into an ideology among scientists in his time. Positivistic and naturalistic interpretations of science have reduced the objective validity of knowledge to subjective strings of appearances and factual inductive generalisations. Husserl claims that such 'positivism, in a manner of speaking, decapitates philosophy' (Husserl 1970a: § 3 p. 9, Hua VI 7). What was the motivation for scientists to espouse positivism? Positivism wants to be loyal to experience but misconceives the nature of that experience. When Husserl latterly became aware of the programme of the Vienna Circle, he regarded it, while a healthy bulwark against the crisis of irrationalism, as nevertheless a flawed philosophy because it had not undergone critical self-interrogation of the manner Husserl required of all philosophy.

In analysing the success of the positive sciences, Husserl is struck by the fact that these successful sciences, when seeking a philosophical elucidation of their nature, are drawn to empiricism as their 'dominant conviction . . . the solely dominant one among empirical investigators' (Husserl 1983: § 18 p. 34, Hua III/I 34). Why should this be so? Why does science feel comfortable with the denial of essences and repudiation of the cognition of essences? Empiricism 'springs from the most praiseworthy motives', but it too carries a conceptual and unexamined baggage (Husserl 1983: § 18 p. 35, Hua III/I 34). Husserl acknowledges that empiricism is 'a radicalism of philosophical practice', setting itself against all idols of superstition, including Scholastic entities such as 'ideas' and 'essences' (Husserl 1983: § 19 p. 35, Hua III/I 35). Empiricists start from 'unclarified preconceived opinions' (Husserl 1983: § 20 p. 38, Hua III/I 38). Husserl believes empiricism must eventually endanger the progress of science as such. For Husserl, empiricism is absurd because the claims it makes are not justified by its own standard of what constitutes meaningful expression. Thus, *avant la lettre*, Husserl was already in possession of the standard criticism of logical positivism, namely, that its criterion of meaningfulness is not in conformity with the conditions laid down by the criterion itself. *Logical Investigations* § 26, 'On Certain Basic Defects of Empiricism', characterises extreme empiricism as just as 'absurd' as scepticism. Indeed Husserl sees empiricism as a kind of scepticism. How does empiricism arrive at its general statements, such as, all meaningful judgements relate to experience? Empiricism puts its trust in singular judgements of experience, yet it justifies its principles and universal laws mediately through induction (Husserl 1970b: *Prolegomena* § 26 p. 116, Hua XVIII A85). What guarantees the truth of these inferences? What principles justify such induction, what principles govern this mediate inference? Empiricists are forced to appeal to psychological regularities in Humean fashion. Empiricism thus confuses the psychological origin of judgements with their validity and becomes a form of psychologism. Incidentally, Husserl absolves his hero Hume of such an absurd radical empiricism; he sees Hume rather as a 'moderate empiricist' who retained logic and mathematics and gave them a priori justification (Husserl 1970b: *Prolegomena* § 26 p. 117, Hua XVIII A86). Empiricism confuses return to things themselves with a demand for the legitimisation of all cognition by *experience* (Husserl 1983: § 19 p. 35, Hua III/I 35). The radical empiricist assumes that the only access to things themselves comes through immediate sensory experience. But, for Husserl, natural things do not constitute the whole set of kinds of things, and thus empiricism at best only reveals things of nature. Husserl claims not all kinds of judgements get their intuitive fulfilments from sensory experience; the empiricists have not understood the whole range of judgements. We cannot simply postulate or dictate in advance the range of forms of judgements and their manner of fulfilment. We can only gather this through 'insight' (Husserl 1983: § 19 p. 36, Hua III/I 36). Immediate seeing is not merely sensuous; it is original presentive (*gebende*) consciousness of any kind. Husserl wants to substitute 'intuition' in a broader sense for 'experience' so that, as he had already seen in *Logical Investigations*, we can have genuine non-sensuous intuitions with the absolute apodictic certainty of eidetic truths, such as the principle of non-contradiction (Husserl 1983: § 20 p. 37, Hua III/I 37).

Perhaps the strongest critique of *naturalism* written in the first half of the twentieth century is Husserl's essay *Philosophy as a Rigorous Science* (1911), commissioned by Heinrich Rickert for his new journal, *Logos*. Husserl saw his era as caught in two post-Hegelian developments. On the one hand, Hegel's notion that every philosophy was an expression of its own time, and hence somehow right for it, had seriously weakened the demand for scientific philosophy, and had eventually led to the development of a sceptical historicism typified by Dilthey (Husserl 1965: 77, Hua XXV 7). Similarly, in reaction to the kind of speculative system of Hegel there developed a renewed interest in Eighteenth-century materialism leading to positivism which gave rise to scepticism. Naturalism does have the ideal of philosophy as a strict science (Husserl 1965: 78, Hua XXV 8). So, for Husserl, 'it is important to engage in a radical criticism of naturalistic philosophy' (ibid.). Positivism, for Husserl, emerges from a naturalised reading of Kant or from Hume (Husserl 1965: 80, Hua XXV 9). For naturalism, physical nature is grasped as a complex of sensations. Indeed naturalism includes the project of 'naturalising' consciousness. Husserl argues that naturalism is self-refuting. In his critique of naturalism, Husserl refers to *Logical Investigations* §§ 25–9 and indeed he always looked back at these sections as an effective philosophical refutation of naturalism and positivism.[13] Similarly, Husserl recognised the positive aspiration of naturalism in that it sets out to achieve philosophy as a rigorous science (Husserl 1965: 78, Hua XXV 8). As such naturalism would always be the most enduring temptation for scientists, but nevertheless, he also criticised its 'naturalistic objectivism' as containing an inbuilt absurdity (*Widersinn*). This absurdity consists in the attempt to naturalise consciousness. Indeed, in the *Crisis*, despite his antipathy to German idealism of a speculative kind, Husserl acknowledged that transcendental idealism was the only philosophy to have successfully resisted the lure of naturalism (Husserl 1970a: 337, Hua VI 271).

The clarification of the scientific ideal

Husserl's first major attempt to clarify the nature of the ideal of science was in the *Logical Investigations* (1900–1). Here, Husserl sharply distinguished between the human methodologies and processes involved in winning scientific insights and the ideal nature and unity of scientific knowledge in itself. As Husserl puts it in the First Logical Investigation § 29:

> All theoretical science consists, in its objective content, of *one* homogeneous stuff: it is an ideal fabric of *meanings* (*eine ideale Komplexion von Bedeutungen*). We can go even further and say that the whole, indefinitely complex web of meanings that we call the theoretical unity of science, falls under the very category that covers all its elements: it is itself a unity of meaning.[14]
>
> (Husserl 1970b: § 29 vol. 1 p. 325, Hua XIX/I A95)

In *Formal and Transcendental Logic* (1929) Husserl speaks in positive terms about science as a closed system of statements. As late as the *Crisis* Husserl still clings to the view that mature science is a single system of interconnecting 'truths in themselves' or 'propositions in themselves', demonstrating that Husserl never completely abandoned Bolzano's conception of science.[15] This led Max Horkheimer in his 1937 lecture 'Traditional and Critical Theory,' a manifesto for the newly founded Frankfurt School, to characterise Husserl as 'traditional' rather than 'critical'; he exemplifies a bourgeois, passive standpoint towards scientific knowledge (Horkheimer 1972). Husserl himself was fully aware of the distinction between the theoretical *ideal* of a domain of fixed truths and the discursive, critical, intersubjective *practices* which humans carry on in order to achieve scientific knowledge. Indeed, it is these latter aspects which came to feature more and more in Husserl's work culminating in his analysis in the *Crisis* and which was strongly influential on Habermas' critique of instrumental reason.[16]

Husserl's conception of scientific *practice* has not received as much attention as it ought, due to Husserl's own heavy emphasis on science as pure *theoria*. But Husserl recognised that scientific practice was constituted by consensus among a *community* of free rational inquirers, 'a community of purely ideal interests' (Husserl 1970a: 287, Hua VI 334). Although Husserl's goal of universal science and of *mathesis universalis* today sounds rather remote, given the diversity of scientific methods which now flourish side by side, his account of scientific practice has a much more contemporary ring. There is no science without humans engaging in co-operative, intersubjective practices and today, in the wake of Wittgenstein, Kuhn, and Feyerabend, on the one hand, and Habermas, Apel and Gadamer, on the other, there is much more interest in how these practices come to be validated.[17]

Actual discoveries must be repeatable (at least in principle), and science modifies its truths over and over again. There is in all scientific endeavour, Husserl claims, both the *ideal* of a convergence towards the truth, and also the recognition that such convergence is the result of human consensus and intersubjective agreement among agents. The key point is that humans recognise the essential truths and are able to carry out and repeat for themselves the insights leading to the scientific discoveries. This communication and iterability is enabled by the use of symbols and written signs which, as it were, strip the personal occasion from utterances and make something in principle intersubjectively graspable. Humans gain mastery over the infinite world of appearances through symbolisation. For Husserl the scientific impulse is quickened by the ability to operate with symbols. The problem is that this symbolic approach to knowledge has not itself been interrogated.

For Husserl, the individual special sciences, even in their success pursuant upon their very emancipation from philosophy, also suffered from a deformation in their development because they had abandoned the philosophically generated *ideal* of genuine science and had naively seized on individual methods and practices as self-justifying. Though technically productive, these practices had never been theoretically interrogated and thus remained 'one-sided', prone to accepting ideologically distorted conceptions of their nature and practice (Husserl 1969: 3–4, Hua XVII 3). Natural science's bracketing of everything subjective leads to a 'bad theory regarding a good procedure' (Husserl 1965: 105, Hua XXV 28).[18] This lack of reflection meant that the European sciences had lost their belief in themselves and in their absolute significance; that is, the sciences in their practices have become utterly divorced from the ideals of a genuinely human way of life (Husserl 1969: 5, Hua XVII 5). They have been reduced from theoretical insight (*Einsicht*) to mere technological practice, a process which Husserl calls the 'technicisation' (*Technisierung*) of method involving an emptying out of meaning (Husserl 1970a: § 9g p. 46, Hua VI 45). In fact, Husserl took the view that the developing formal logic of his day (which he called *Logistik*) suffered from being merely a theoretically 'naive' technology rather than a fully transparent theoretical practice since its grounding in the life-world was unclear (Husserl 1970a: § 36 p. 141, Hua VI 144).[19] Husserl remained unconvinced of the philosophical advantages of mathematical logic for the clarification of thinking because he could not see the relation between a calculus or set of algorithms and the theoretical insights which would justify them. But, as Herbert Marcuse points out, the inherently instrumental character of science is something which Husserl diagnosed as coming before any technical application, and to be due to the process of symbolisation itself (Marcuse 1965, especially p. 286).

In the face of this lack of rational self-reflection, Husserl's whole philosophical career was motivated by the project of *clarification* of the grounding concepts of the sciences and philosophy. In a private diary from 1906 he remarks 'I simply cannot go on without clarity (*Klarheït*). I will – I must – approach these sublime goals, through self-sacrificing labor and purely disinterested absorption in the work. I am fighting for my life, and because of this have confidence that I will be able to make progress. . . . Only one thing will fulfil me: I must come to clarity!' (Husserl 1994a: 494 Hua XXIV 445). This Holy Grail of 'clarification' was understood by him as a challenge to make all human life and action, including our entire commitment to theoretical knowledge, transparently rational and self-consciously affirmed as such. To achieve this clarification nothing less that a thorough-going critique of normative reason was required, a critique of logic and knowledge and of the whole sphere of human awareness, including not just cognition but all our pre-cognitive commitments including our spatial, temporal and bodily awarenesses as well as our insertion into a flowing unified, conscious, emotional life, both individually experienced and understood from the viewpoint of culture and the life of 'spirit'.

For Husserl this clarification came only from a *general phenomenology* of knowledge and consciousness, from what he called 'radical investigations of sense' (*Besinnungen*), deliberations which attempt to grasp and theoretically reconstruct the 'sense fulfilments' (*Sinnerfüllungen*) constitutive of knowledge in its highest form. All genuine knowing, for Husserl, consists of a kind of evident cognition or self-evident insight. Thus, in perception, a cognitive act of perceiving is evident if the object is present in full bodily givenness (*Leibhaftigkeit*) and this is recognised to be so. Phenomenology, as the investigation of the structural characteristics of intentionality, was to provide an account of different kinds of intuitive fulfilment, how 'objectivities' (*Gegenständlichkeiten*) come to be framed in the different disciplines. For Husserl, phenomenology must always proceed through winning insights and was never to a deductive system based on consequential logic.

The constitution of objectivity

After the *Logical Investigations*, Husserl's interests broadened beyond the clarification of logic and mathematics to an attempt at a general theory of knowledge as such, including all normative knowledge, the foundations of value, and so on. This 'constitutive' phenomenological inquiry eventually needed to be complemented by an 'archaeology' of the history of cultural development, which he called 'genetic phenomenology'. But, even in this attempt in his late writings to locate science within history, Husserl is not surrendering to an anti-scientific historicism whereby scientific achievement is considered merely as the expression of a world view or *Weltanschauung*. Quite the opposite. Husserl wanted to rescue philosophy from mere changing 'world views'. He was suspicious of the neo-Hegelian 'sceptical historicism' of the cultural sciences, espoused by Dilthey and others, seeing in it a kind of creeping relativism which would inevitably lead to moral nihilism and irrationality. He saw the need for a questioning back; first to uncover pre-scientific life and then to uncover transcendental subjectivity. But as for the phenomenal progress of the individual sciences in themselves, he says in *Ideas I* (1913): 'When it is actually natural science that speaks, we listen gladly and as disciples' (Husserl 1983: § 20 p. 39, 1950: III/I 38). But often it is not natural science in a pure form which 'speaks' rather some form of prejudice is being articulated. This is especially true when scientists talk about their own work, they often assume a current ideology (e.g., positivism in Husserl's day, or perhaps, the manner in which Popperianism is a dogma among practising scientists today).

Like Kant, Husserl himself entertained no sceptical doubts about the possibility of objective knowledge. Rather his problem was to understand how this objectivity is 'constituted' in and through the acts of subjects. Philosophy must inquire into the 'subjective conditions of the possibility of an objectively experienceable and knowable world' (Husserl 1970a: § 29 p. 112, Hua VI 114). Indeed, both his early and his late texts – *On the Concept of Number; Psychological Analyses* (1887) and *Origin of Geometry* (1936) – show fascination with the same problem, the single enduring problem of his philosophy. This for Husserl is 'the miracle', 'the mystery of mysteries', 'the enigma' of philosophy. Indeed he is puzzled that traditional philosophy has been so lax in studying this area, thereby allowing a deficient naturalism to claim that the encounter with objectivity is a factual physical process. For Husserl, objectivity involves notions of self-identity, continuity over time (even atemporality), and universality: features which he accepts, with Hume and Kant, are not given in sensuous intuition.

In opposition to all forms of sensualism and atomistic empiricism, Husserl accepts that in acts of intuiting (perceiving, remembering, imagining, knowing) essences are presented. This is Husserl's central conception of *Wesensschau*, the viewing or inspection of essences. To understand the nature of objectivity one had to get over the 'Humean confusion' whereby it is thought that in intuition one grasps not the essence but mere factual details corresponding to essences (Husserl 1965: 115, Hua XXV 36). Initially, in the *Logical Investigations* he talked about the manner in which we have a 'categorial intuition' of these idealities, and later situated these as transcendental structures of the transcendental ego. Here, the manner of his relation to Descartes and Kant becomes crucial. Leaving aside the difficult and problematical issues surrounding Husserl's positing of a transcendental ego, Husserl's central concern was that science should be concerned with essences, with a priori universal, and hence necessary, knowledge. For Husserl, justification depends on on-going acts of insight or evidence, acts in which matters are seen just as they are.[20]

Already in his *Habilitationschrift* written under Carl Stumpf, Husserl had attempted to explain the origin of mathematical objects in subjective acts. He employed Brentanian descriptive psychology to explain the origins of the number concepts in our conscious acts. How do the concepts of number arise for us? His answer differs significantly from the traditional empiricist account given by Mill and others. The number concepts, for Husserl, are specifications and differentiations between concepts of multiplicities. Husserl maintained, as Gestalt psychologists such as Ehrenfels did, that we see not just individuals, but groups and collectivities. Selecting the items we will include in a specific group depends on our interests and is not a purely passive experience. Collective combination of items in a group is a matter of seeing relations of a special type. In order to form concepts of multiplicity, we abstract from the individual properties of the items and treat each member of the group merely as a 'something' (*etwas*). That is to say, invoking a version of Brentano's distinction between physical and psychical relations, we grasp the items not on the basis of any 'physical' or 'content' reations between them but solely on the basis of making a psychic connection. The notion of number is based on the ability to relate together objects from different categories simply by abstracting from their specific characteristics and treating them as unities. Thus, we can see that a pen, an apple and a painting as a multiplicity of 'a something and a something and a something', from which the number 3 derives. This seems to suggest that numbers are properties of groups, for Husserl. For Frege, on the other hand, numbers are not properties of groups of objects, but *extensions* of concepts. Diverse entities can only be counted together if they are brought under a concept, e.g., the number of *cities*. Husserl's account of numbers, Frege says in his sharp review of the *Philosophy of Arithmetic*, since it confused the objective validity of logical and mathematical truths with their modes of givenness is consciousness, amounted to *psychologism*.

Though Husserl acknowledged the force of Frege's criticisms, he had already moved beyond the psychologistic elements in the *Philosophy of Arithmetic* as is evident from his 1891 critique of Schröder's *Algebra der Logik* (a pre-Fregean German attempt at mathematical logic) and from his unpublished review of Brentano's Polish student, Kasimir Twardowski's *On the Content and Object of Presentations* (Husserl 1994: 52–92, 388–95). Furthermore, in his private correspondence with Frege, and also because of his reading of Lotze and Bolzano, two logicians who also influenced Frege, Husserl was already clearly distinguishing between the psychological act and its objective (or ideal) content. In other words, Husserl was already developing a theory of sense (*Sinn*), later to become the theory of the noema in *Ideas I* (1913). He differs from Frege in seeing *Sinne* in all intentional acts, not just in linguistic sentences. There is a perceptual *Sinn* in seeing, in remembering and so on. As a matter of fact, Husserl never completely rejected the *Philosophy of Arithmetic* in later life and came to see it as constitutive analysis which had phenomenological aspects rather than being a psychologistic treatise. After all, it was a central insight of Husserl's all through his life that ideal meanings (such as *the square root of 9*) signify independent ideal objectivities which nevertheless are brought to appearance in human, temporal, finite acts of consciousness. Furthermore, Husserl's recognition that human thinking can only come to grips with the infinitely large domain of numbers by moving to symbolisation, a position expressed in the *Philosophy of Arithmetic*, was to become a foundation stone for his analysis of both the success and the distorting impact of the mathematical sciences of nature in the modern period.

After 1894 Husserl's task was to see a way of grounding human scientific knowledge by exploring the nature of the relation between acts of consciousness and the domain of ideal meanings. The first outcome of this exploration was the huge *Logical Investigations* of 1900–1 which aims at the 'epistemological criticism and the clarification of pure logic' (Husserl 1970b: vol. 1 p. 249, Hua XIX/IA3). Here Husserl laid to rest the ghost of his earlier psychologism and undertook a number of studies of elements in logic and semantics, including an attempt at a phenomenological account of signification and signs (not unlike Peirce). The task of the *Logical Investigations* was to trace the manner in which these objectivities (*Gegenständlichkeiten*) are constituted. The *Prolegomena* to the *Logical Investigations* is a sustained critique of psychologism and naturalism. Thus for Husserl, Newton's law of universal gravitation is true whether or not anyone ever discovers it. This led Husserl to further studies in logic and epistemology and the announcement of the project of phenomenology, understood at this time as the study of essences.

According to the *Prolegomena*, the dominant fashion of nineteenth-century theorists of science, had conceived of the discoveries of science as mere factual, empirical generalisations. Husserl, on the other hand, conceived of sciences as a fixed, unified domain of ideal theoretical truths, interconnected by ideal laws. Such was logic and mathematics, and such as he conceived it, would be physics and the other sciences.[21] Quite separate from this domain of ideal laws, were the various technical disciplines (*Kunstlehre*) which applied these ideal laws as norms for the regulation of behaviour. But in the *Logical Investigations*, Husserl, influenced primarily by Lotze's interpretation of the Platonic forms and applying Lotze's view to the interpretation of Bolzano's 'propositions in themselves' (*Sätze an sich*), sharply distinguished between the domain of ideal theoretical truths and the behaviour-guiding norms which derive from them.[22] To collapse one into the other would lead to the danger of psychologism. This is precisely what happens when an ideal objective truth, e.g., the Principle of Non Contradiction, is interpreted as merely a norm guiding human thinking or as a psychological, empirical law governing the association between actual thought processes. Right through to the *Crisis* Husserl was emphatic about the need to differentiate between a normative practice, what the Greeks called *techne*, and genuine theoretical knowledge (*episteme*). Husserl wanted to preserve the domain of the ideal as a genuine domain of which we can have scientific knowledge without relapsing into Platonism. Ideal objectivities are constituted through repeated acts and are grasped as the self-same in those acts. But they are not to be thought of as independently existing objects in a Platonic sense, they are abstract, ideal entities which underwrite 'unities of meaning' (*Bedeutungseinheiten*). Clearly to believe that we can have knowledge of the 'ideal' is this sense means that Husserl may be construed as an idealist, but he rejected the label of 'Platonism' for his view, as much as he denied that he was a subjective idealist about the nature of ideal truths. He seemed to regard Platonism as committed to a belief in the immaterial existence of these ideal objectivities. Husserl's specific contribution was to recognise the multiplicity of kinds of ideal objects, mathematical, musical, aesthetic, and so on. For example, in aesthetics, Husserl distinguished between the kind of ideal meaning which is bound to a single unique artwork (Raphael's *Madonna*) and those which can be instantiated over and over again (e.g., Goethe's *Faust*) (Husserl 1973a: § 65 p. 266).

In fact, it was his concentration on analysing the structures in our knowing process which guarantee the objectivity of our knowledge which led many of his critics – including Heidegger – to assume that Husserl had collapsed back into psychologism in the second volume of the *Logical Investigations*. But Husserl's mature notion of *constitution* (*Konstitution*) is precisely distinguished from merely factual, empirical psychological processes. Husserl leaned more and more towards Kantian formulations which identified these constitutive structures as transcendental, that is, as conditions for the possibility of knowledge in general (*überhaupt*), though Husserl had a much broader sense than Kant of what these conditions might be. Ultimately, this led Husserl to positing a transcendental ego – not to be understood as an empirical ego writ large – as a unified condition for the possibility of objectifying structures in consciousness. By the time we get to *Ideas I* (1913), Husserl was articulating a project of a pure science of consciousness, construed in a priori terms. Husserl had been reading Kant since the 1890s but began seriously to orient himself in Kantian idealist manner after 1905. For Husserl, the a priori science of pure consciousness is a bedrock science since all other forms of knowing presuppose and utilise the very structures of consciousness itself. Husserl came to recognise that, in our ordinary relations to the world, these structures of consciousness are presupposed, and in a sense masked, and he realised that a special exercise of attention, what he called 'phenomenological bracketing' (*epoché; Einklammerung*), or putting in parenthesis, and various forms of 'reduction' (philosophical, phenomenological, transcendental, eidetic reductions) were necessary to lay bare the a priori structures of consciousness. Husserl spent a great deal of his published works analysing how 'the natural attitude' (*die natürliche Einstellung*) needed to be suspended in order to explore the realm of ideal essences which is consciousness. Husserl was impressed by the fact that an eidetic science like geometry was able to move seamlessly from the factually given space to ideal space, whereas, in the investigation of consciousness, there is no smooth transition from the ordinary 'Heraclitean flux' of conscious life to the ideal insights concerning the essential nature of conscious forms. Hence the problem of reduction came to dominate his thinking as Husserl tried to disentangle the natural urges to objectify and to posit as real from the more neutral description of acts generating knowledge.

The role of the lived body

Somewhat paradoxically, at the same time as Husserl was laying out his transcendental idealism in *Ideas I,* he was also exploring a different direction, namely the manner the human body shapes the nature of conscious awareness and installs that awareness in the world. The living human body with its eyes and specific sense organs, located as they are with its range of motor movements and nerve endings, restricts and structures our experience in a manner which had not been adequately articulated by the prevailing mechanistic physiology and behavioural psychology. Husserl's analysis of the distinction between the incarnate animate body (*Leib*) and the inanimate physical body (*Körper*), subject to physical laws, as developed in the posthumously published *Ideas II* and as later revisited in the *Crisis of European Sciences*, had a huge impact on both Heidegger's account of being-in-the-world in *Being and Time* (1927) and on Merleau-Ponty's account of the body-subject in his *Phenomenology of Perception* (1945). The lived body experiences the world as an environment (*Umwelt*).

Husserl's discussion of the animate body (*Leib*) was groundbreaking and still presents a strong challenge to materialist and reductionist models of the body operative in conventional medicine and psychology. Husserl felt the need for a rigorous investigation which would reconnect the idealities of geometrical space to the experiential space of our lived experience. This led to his desire to see the world and our human involvements under a different eye, no longer in the natural attitude which itself was closely tied to naturalism in physics. Where do our concepts of space and time come from? How do the scientifically purified forms of these concepts relate to our pre-scientific worldly conceptions of time and space? The advances in the mathematicisation of nature had led to ideal properties of space and of objects been seen as the 'real' properties whereas the phenomenological properties belonging to our everyday encounter with the world were somehow treated as secondary characteristics, if not as wholly illusory.

On the other hand, Husserl recognised that our bodily insertion into a spatial world was the source of a very particular and complex experience of lived space, one which had not been conceptualised. In a sense, the body is the locus of all reference, the zero-point of perceptual acts. This pre-conceptual lived space is neither Euclidean nor anti-Euclidean, but is expressed in bodily orientation, left and right, up and down, the upright posture, the experience of our bodily weight and resistance to movement, and the various forms of motility of our bodily organs. Phenomenology needs to ground scientific conception of space and time in this bodily-based lived field of experience. Thus, for example, the manner I possess my body needs to be carefully studied. To use an example which would appeal to Merleau-Ponty, the anorexic will see herself as fat even though she can recognise the look of a starving body as shown in photographs. The scientific image of the body needed to be supplemented by the lived image. To be complete, science must reconnect itself with the ground from which it first emerged, and from which, in order to develop its unique method of abstraction and symbolisation, it had to cut itself off. In part, this reconnection of science and lived experience required recognising that the processes of the objectification of meaning had a temporal or historical dimension. Husserl became interested in the 'genetic' aspect of the constitution of meaning side by side with the static model he had earlier proposed. Husserl sees the understanding of the genesis of the idealisation of science and the manner in which that has been distorted by positivism and naturalism as opening the possibility of seeing new opportunities for developing the scientific ideal in a non-distorting manner.

The Crisis of European Sciences

In his 1936 work *The Crisis of European Sciences* (only the first part was published in 1936; the manuscripts left unfinished at his death were published in full posthumously), Husserl diagnosed a general crisis evident in the sciences as a whole, including in mathematics, always considered as the model of what a science ought to be. Husserl understands a scientific 'crisis' as occurring when the manner in which that science sets its task and method becomes questionable (Husserl 1970a: § 1 p. 3, Hua VI 1).[23] Husserl's notion of crisis, then, is unrelated to the actual success of a science. Indeed, Husserl has enormous respect for the achievements of contemporary physics in particular. Rather, the crisis of science occurs when the relation of science to its philosophical goal becomes problematic, and hence the meaning science has for human existence (*menschliches Dasein*) becomes doubtful (Husserl 1970a: § 5 p. 12, Hua VI 10). According to Husserl, the nineteenth century allowed itself to be blinded by the 'prosperity' of the positive sciences. The success of the fact-minded sciences produced a generation of fact-minded people who excluded from science all questions of human existence as a 'free, self-determining being' (Husserl 1970a: § 2 p. 6, Hua VI 4). The physical sciences' relentless quest for objectivity has led to the exclusion of everything subjective. No attention has been paid to the manner that exact objectivity arose as an ideal. Mathematical objectivity has been a transformative notion which utterly changed our relation to the world, but itself has not been interrogated. How do we move from the Heraclitean flux of individual experience, to communal mutual confirmation, to an ideal of the objective as something standing entirely independent of us? In particular, the very possibility of a science's accomplishment of objectivity has become problematic (Husserl 1970a: § 33 p. 122, Hua VI 124).

The nature of this ideal of objectivity is puzzling, the story of its genesis is complex. But it is this inquiry which Husserl seeks to carry out in the *Crisis*. Husserl proposes a kind of genetic phenomenology (what Foucault would call 'archaeology') to understand how science has come to shape our world view. He wants an inquiry into how the pre-given life-world gives rise to and provides the 'subsoil' (*Untergrund*) for the discovery of theoretical truths (Husserl 1970a: § 34 p. 124, Hua VI 127). This investigation is not empirical factual history in the usual sense, but rather, a kind of intellectual reconstruction, 'a teleological-historical reflection upon the origins of our critical scientific situation' (Husserl 1970a: § 1 p. 3, Hua VI 1).

> We shall attempt to strike through the crust of the externalized 'historical facts' of philosophical history, interrogating, exhibiting, and testing their inner meaning and hidden teleology. Gradually . . . possibilities for a complete reorientation of view will make themselves felt, pointing to new dimensions.
>
> (Husserl 1970a: § 7 p. 18, Hua VI 16)

We should not expect this intellectual reconstruction to be completely factually accurate; and Husserl has been accused of making Galileo stand for positions which should more accurately be ascribed to Descartes. But Husserl is painting a picture, showing the pattern of thinking at work, getting to what he considers to be the *essence* of modernity's conception of science: 'Our concern is to achieve complete clarity of the idea and task of a physics which in its Galilean form originally determined modern philosophy' (Husserl 1970a: § 9e p. 42, Hua VI 42). In this sense, all modern science is Galilean in its mathematical conception of nature; Einsteinian physics, for Husserl, is part of 'Galilean science'.

Galilean science has idealised and mathematicised nature, leading to an abstraction from the lived, experienced world. Thus, modern science approaches nature not as a complex of individuals of infinite shadings and complexities, but conceives of 'nature as idea, as regulative ideal norm, as the logos, in a higher sense, belonging to actually experienced nature' (Husserl 1969: 292–3, Hua XVII 257).[24] Gradually this idealised normative conception of nature has replaced the inexact continuum of our sensory experience. The more successful the science, the more it has engaged informalisation and the 'emptying out of meaning' (*Sinnentleerung*) through relentless symbolisation. Galileo is 'the creator of the conception which first made physics possible' by taking for granted the universal applicability of mathematics (Husserl 1970a: § 9 pp. 36–8, Hua VI 35–7).[25] For Galileo the book of nature is written in the language of mathematics. Nature has now been understood as a 'mathematical manifold' (Husserl 1970a: § 9 p. 23, Hua VI 20). In Galilean, and hence in all modern, science, the key to the success has been to abstract from the particularities of bodies and treat them as ideal geometric shapes obeying ideally determined exact laws. Science is less interested in the empirical fact than in the formulation of ideal laws. Mathematics thus idealises the world of bodies: 'One can truly say that the idea of nature as a really self-enclosed world of bodies first emerges with Galileo' (Husserl 1970a: § 10 p. 60, Hua VI 61). For instance, in measuring falling bodies, we can abstract from their irregularities and treat them simply as centres of mass governed by the law of gravity. First there is the geometricisation of nature, and then geometry itself is construed in terms of algebra (in Descartes, Vieta and Leibniz). Shapes are transformed into purely numeric configurations. Eventually, science has replaced the experiential world completely; the world of human experience has even been assigned the value of illusion and mere appearance. All 'subjective-relative' properties, such as colour, taste, and the other so-called 'secondary qualities', have been dispensed with. The mathematically ideal world has been 'substituted' for the real world, and the mathematical garb of symbols (*Ideenkleid*) has been mistaken for the real objective world (Husserl 1970a: § 9h pp. 48–51, Hua VI 49 ff.).

But modern science does not merely bring about a divorce between the lived world and the world as described by mathematical science. The formalisation of nature also leads to a radical alteration in the nature of individual subjectivity. For Husserl, when a group of scientists record observations, they assume a certain *substitutability* between one observing subject and another. One human can come to stand for another. The 'subjective relative' aspect of human experience is put entirely out of account; the difference in perspectives between subjects is ignored or bracketed. When scientists think that they encounter the objective world they forget the manner in which this objectivity is ideal and constructed and is akin to experiencing of the infinite number series:

> The empiricist talk of natural scientists often, if not for the most part, gives the impression that the natural sciences are based on the experience of objective nature. . . . The experienceability of something objective is no different from that of an infinitely distant geometrical construct and in general no different from that of all infinite 'ideas', including, for example, the infinity of the number series.
>
> (Husserl 1970a: § 34d 128–9, Hua VI 131–2)

The point, for Husserl, is that physical objects are experienced in the world from a multiplicity of perspectives, but they can never be grasped all at once, which is the manner of cognising an abstract entity. The specifically human way of perceiving and engaging with the world has been excluded. Alternatively, when it is included, it is relegated to another natural science, psychology. According to Husserl's sketch of the inner working out of the conception of modem mathematical science, the splitting of the subjective-relative from the mathematical objective world inevitably led to the dualism of matter and mind found in Descartes, and thence to the conception of psychology as a split-off separate science modelled on natural science, articulated first in Locke and Hume but still prevalent (Husserl 1970a: § 22 84, Hua VI 86). Husserl correctly diagnoses that our ordinary world view has now been affected by this scientific outlook, that in our ordinary language and attitudes we reflect these scientific presumptions.

Although Husserl had identified the role of our cultural environment (*Umwelt*) in shaping our approach to knowledge, as early as *Ideas* II, it was not until the *Crisis* that he focused more specifically on what he called the 'pre-scientifically intuited nature' (Husserl 1970a: § 9h p. 50, Hua VI 50). The pre-given life world is the 'grounding soil' (*der gründende Boden*) for the scientifically true world (Husserl 1970a: § 34e p. 131, Hua VI 134). The historical and cultural life world has determined the shape of science, yet modern philosophy has forgotten the founding relation between our 'pregiven life work' and the scientific outlook; this can only lead to distortion, a 'sliding over' or 'concealing of meaning' (*Sinnüberschiebung*, *Sinnüberdeckung*), threatening scepticism. To restore the balance, Husserl wants to focus on the original 'bestowal of meaning' (*Sinngebung*) which enabled the formal system of mathematical science to be in some useful sense *about* the world (Husserl 1970a: § 9g p. 47, Hua VI 46). The genesis of 'exact objectivity' as an ideal is a specific human 'cognitive accomplishment (*eine Erkenntnisleistung*) (Husserl 1970a: 347, Hua VI 360). Husserl recognises this 'rational and all conclusive' goal of rationality as the beginning of science proper.

Husserl is very interested in the moment when a science releases itself from local considerations, e.g., geometry's beginning in measuring fields or areas, and recognises its universal and infinite task: 'Not until the dawn of the modern period does the actual discovery and conquest of the infinite mathematical horizons begin' (Husserl 1970a: § 8 p. 22, Hua VI 19). All through his life Husserl was deeply interested the relation between geometry and naturally perceived or intuited, lived space. Husserl had proposed to study the nature of geometry in his earliest investigations into the foundations of mathematics. In later years, his inquiry took the form of trying to understand the fateful turn taken in the seventeenth century with the geometricisation of nature and of reason in Descartes, Spinoza and others. Ancient geometry began in land surveying, and this original giving of meaning enabled the ideal mathematical vision to be accomplished. Unfortunately, Galileo never reflected on this move and hence there is the illusion that geometry as independent sphere of self-sufficient truth could be applied willy-nilly, as in the attempts by Descartes and Spinoza to found all knowledge as a deductive system from evident truths on the model of geometry (*more geometrico*). Husserl regarded this as a failure because the starting point was naive. The ideal of objectivity began in mathematics but was transformed into a new tool in the idea of 'mathematical natural science' in Galileo (Husserl 1970a: § 8 p. 23, Hua VI 20) and if the ideal has been accomplished it is precisely in mathematics and mathematical physics (Husserl 1970a: 347, Hua VI 360).

Evaluating Husserl's contribution

How should we characterise Husserl's engagement with and critique of the sciences, especially their impact on human culture? On the one hand, Husserl is a defender of science and the objectivity of science. He always retains the view that science is driven by an ideal of objective, universal truth and that, as such, there must be unity of the domain of scientific knowledge as a whole. However, though he was familiar with Hilbert's project of formal axiomatisation, Husserl's own vision of science emphasised the need to connect the system of truths with the acts of intellectual cognition and insight which gave birth to them. He, therefore, soon came to doubt that pure logic or mathematics could provide a fundamental basis for all forms of science. Rather, for Husserl, the ideal of science and its achievements can only be understood when the subjective acts giving rise to the scientific outlook are themselves examined and clarified as to their nature, and when their subjective and cultural specificities are taken into account. Phenomenology, for Husserl, was precisely the dream of a science which would keep the guiding ideal of rationality operative in the sciences secured in the clarification of the fundamental meaning-constituting acts of human subjectivity and intersubjectivity.

Husserl has been criticised for not clearly explicating the relation between the *Lebenswelt* and the theoretical attitude of modern science. Husserl never exactly spells out the relation between the *Lebenswelt* and the scientific frame of reference. He has, for example, been criticised for assuming that scientific concepts are 'grounded' in, and gain their meaning from everyday conceptions, whereas some would argue that scientific concepts over time in fact sediment into the everyday ones (thus we talk of our blood sugar being low when we are tired, and so on). Husserl has also been accused of reintroducing relativism into his picture, since he held that diverse human communities may inhabit different life-worlds, leading to the possibility of different forms of everydayness, different kinds of *Lebenswelt*. But Husserl really wanted to overcome this relativist threat by seeking the invariant features belonging to the essence of life-world itself, invariant features which are there prior to the different particular forms the cultural environment could take in different kinds of societies (thus his interest in Lucien Lévy-Bruhl's description of the primitive mentality, for example).

In a sense, then, Husserl is a *foundationalist*, though he did not agree either with the foundational attempts of empiricism which privileged sense data, nor with the rational foundations proposed by Descartes. Husserl's foundationalism holds that all knowledge is ultimately justified by self-evident insight. But these evident insights are actually discoverable in many different kinds of acts, and the conditions of satisfaction for these evidential acts vary depending on the kind of knowledge involved. Husserl is clearly aware, however, that this self-evidence is an ideal limit to which all knowledge merely approximates. Furthermore, Husserl's inquiry into the nature of the subjectivity grounding knowledge eventually led him to intersubjective and historical inquiries about the shape of Western scientific outlook in general and the presuppositions upon which it rests. Here Husserl can be seen as identifying the theoretical problem which arises when scientific theoretical insights are translated into technological rationality. In a sense, then, it is Husserl who initiated the worries about the global entrenchment of technological reason which one finds expressed in Heidegger, in Marcuse, in the Frankfurt school and in Habermas.[26]

As we have seen, Husserl's critique of inadequate conceptions of science moved in two different directions, attacking both *scientism* and *cultural relativism*. Husserl's efforts to overcome these threats are not dissimilar to Hilary Putnam's project of humanising philosophy and science.[27] Both seek to understand the deep motives for science's lost role in the humanisation of society. Indeed, Putnam acknowledges the later Husserl's conception of the *Lebenswelt* as an important concept in articulating his own attempt to overcome the false dichotomy Putnam has diagnosed in contemporary philosophy between the 'furniture of the universe', on the one hand, and our projections, on the other (Putnam 1990: 50).[28] For Putnam, as for Husserl, science lost its leadership in the domain of cultural values when it lost its connection with the experiential world. But Husserl never wanted to oppose science itself. Rather he wanted to radicalise the very thinking about science, and in so doing draw science back into philosophy.

Notes

1 Sokal and Bricmont (1997) have shown the deficiences in scientific awareness of many contemporary French philosophers. It would be wrong, however, to implicate other European philosophers such as Husserl, Heidegger, Merleau-Ponty or Oscar Becker, in this ignorance. Both Husserl and Heidegger had considerable grounding in science. Though Heidegger went on notoriously to claim that 'science does not think', he is referring there to a specific kind of radical questioning which he considers belongs to philosophy. Post-Heideggerian philosophers, however, have simply ignored the achievements of the exact sciences.

2 See the excellent study by M. Kusch (1995).

3 See Husserl, *Logical Investigations* Prolegomena §71; Hua XVIII A252–55. I shall quote from the English translation of Findlay (Husserl 1970b). References to the German editions are to *Husserliana* (abbreviated hereafter as Hua) followed by volume number and page number (e.g. Hua VI 143).

4 See, for example, Albertazzi, Libardi and Poli 1996 and Rollinger 1996.

5 On Husserl's relation to Cantor, see Hill 1997.

6 Husserl refers directly to Hilbert in Husserl 1969a: 96–7, Hua XVII 84–5, and indirectly to his programme in Husserl 1970a: §9f: 45, Hua VI 44–5. Husserl locates its origin in the Euclidean ideal and also in Leibniz's conception of *mathesis universalis*. On Husserl's relation to Hilbert, see Mahnke 1977 and Heelan 1989.

7 Husserl never departed from this conception of an achieved science in *Formal and Transcendental Logic* (1929). Husserl's view of science as a complete axiomatic system of interconnected propositions is of course challeneged by Gödel's Incompleteness Theorem (1931) which shown the impossibility of Hilbert's aims at complete formalisation. On the other hand, Husserl may be defended on the grounds that he treated this merely as an ideal, and in fact acknowledged the infinite nature of scientific discovery left a great deal of room for different ways of organising the system of propositions. Husserl held that there would inevitably be material forms of relation and dependency which resisted formal axiomatisation. Thus Husserl talks of intuitive unifications which cannot be expressed in the logical calculus in *Experience and Judgment*

(1973a), § 62: 248. For a discussion of Husserl's conception of science as formal axiomatisation in relation to Gödel, see Bachelard 1968: 52 ff.

9 This is from an entry in Husserl's notebook dating from September 1906.

10 Husserl's recognition that theoretical knowing is itself a kind of *praxis* was developed by Heidegger into a central theme of *Being and Time* (1927).

11 Throughout his life Husserl refers frequently to the formulations 'truth in itself', 'propositions in themselves' and so on, usually deliberately evoking Bolzano's project of the theory of science as a theory of objective meanings.

12 Husserl gives more detail on this in his *Erste Philosophie* lectures of 1923–4, Hua VII, and in Husserl 1969a, Hua XVII.

13 Husserl refers to *Logical Investigations* in Husserl 1965: 80, Hua XXV 9, and again in Husserl 1983, § 20: 37–8, Hua III/I 37–8.

14 Husserl's conception here can be compared with Quine's notion of a 'web of belief', see Smith and Woodruff Smith 1995: 33.

15 See for example Husserl 1970a: § 34e: 129–30, Hua VI 132–3. Bolzano's conception of science is set out in *Wissenschaftslehre* §1, translated as *Theory of Science* (Bolzano 1972: 1). Bolzano characterises the theory of science as the manner in which the various *treatises* or textbooks of the sciences are to be laid out and ordered. Husserl departs from this particular interpretation of *Wissenschaftslehre* in Husserl 1970b, *Prolegomena* § 12: 73, Hua XVII A29.

16 Furthermore, Husserl encouraged student like Arnold Metzger who wanted to promote a phenomenology of revolution, for example.

17 See Husserl's 1936 essay 'The Origin of Geometry,' Husserl 1970a: 353–78, Hua VI 365–86, for an analysis of how geometry as a science is made possible by intersubjective communal practices and especially 'by means of language through which it achieves, so to speak, its linguistic living body (*Sprachlieb*)' (Husserl 1970a: 358, Hua VI 369).

18 In his 1925 lectures on *Phenomenological Psychology* (Husserl 1977), Husserl states that the objectivity of modern science is in fact 'an artificial product of method' (Husserl 1977: §5 pp. 39–40, Hua IX 54).

19 See Husserl 1970a: § 36, p. 141, Hua VI 144. For a discussion of Husserl's criticisim of formal logic, see Willard 1979.

20 For a clear discussion of this point see Hanna 1993.

21 For the similarity between Husserl's conception of science and that of David Hilbert and Felix Klein, see Heelan 1989.

22 Frege propsed a similar distinction in his *Grundgesetze der Arithmetik* (1893), translated as *Foundations of Arithmetic,* whereby logical laws are descriptive of ideal truths which can of course be reformulated as norms. See Kusch 1995: 34.

23 In a sense Husserl is carrying out a critique of the metaphysical assumptions of science, also found in Alexandre Koyré and in E. A. Burtt.

24 Husserl has been criticised for overemphasising Galileo's Platonism. Whereas traditional Platonists held that the world of experience cannot be forced into the straitjacket of exact quantification, Galileo believes an exact mathematical pattern is concealed beneath the sensible cloak. See Gutting1978–9.

25 It is noteworthy that historical investigation into Galileo's contribution to science has been carried out by Alexandres Koyré (1892–1964), who studied with Edmond Husserl at Göttingen from 1910 to 1912, in Koyré 1978. Husserl may have been influenced also by the discussion of Galileo in Ernst Cassirer's *Das Erkenntnisproblem*.

26 Heidegger's critique of the global framework of technological reason as expressed in his essay 'The question concerning technology' (Heidegger

1978) is a clear development of Husserl's concerns.

27 See especially Putnam's Carus lectures, *The Many Faces of Realism* (Putnam 1987).

28 Putnam groups Husserl here with the later Wittgenstein and with Austin, see also Putnam 1990 p. 89.

Bibliography

Albertazzi, L., Libardi, M. and Poli, R. (eds) (1996) *The School of Franz Brentano*, Dordrecht: Kluwer.

Bachelard, S. (1968) A *Study of Husserl's Formal and Transcendental Logic*, trans. L. Embree, Evanston: Northwestern University Press.

Bolzano, B. (1972) *Theory of Science*, trans. R. George, Oxford: Blackwell.

Gutting, G. (1978–9) 'Husserl and scientific realism', *Philosophy and Phenomenological Research* 39: 42–56.

Hanna, R. (1993) 'Logical cognition: Husserl's Prolegomena and the truth in psychologism', *Philosophy and Phenomenological Research* 53: 2 (June): 251–75.

Heelan, P. A. (1989) 'Husserl's philosophy of science', in J. N. Mohanty and W. R. McKenna (eds) *Husserl's Phenomenology. A Textbook,* Washington, D.C.: Center for Advance Research in Phenomenology: 387–428.

Heidegger, M. (1978) 'The question concerning technology', in D. Farrell Krell (ed.) *Heidegger. Basic Writings,* London: Routledge.

Hill, C. O. (1997) 'Did Georg Cantor influence Edmund Husserl?' *Synthese* 113,1: 145–70.

Horkheimer, M. (1972) 'Traditional and critical theory', trans. M. J. O'Connell, in *Critical Theory. Selected Essays*, New York: Seabury: 188–243.

Husserl, E. (1950–) *Gesammelte Werke*, Dordrecht: Kluwer.

Husserl, E. (1965) 'Philosophy as a rigorous science', trans. Q. Lauer, in *Phenomenology and the Crisis of Philosophy*, New York: Harper and Row.

Husserl, E. (1969) *Formal and Transcendental Logic*, trans. Dorion Cairns, The Hague: Nijhoff.

Husserl, E. (1970a) *The Crisis of European Sciences and Transcendental Phenomenology*, trans. D. Carr, Evanston: Northwestern University Press.

Husserl, E. (1970b) *Logical Investigations*, trans. J. Findlay, 2 vols., London: Routledge and Kegan Paul.

Husserl, E. (1973a) *Experience and Judgment,* trans. J. Churchill and K. Ameriks, Evanston: Northwestern University Press.

Husserl, E. (1973b) *Zur Phänomenologie der Intersubjektivität: Texte auf dem Nachlas, Erster Teil*: 1905–1920, ed. I. Kern, Dorderecht: Kluwer.

Husserl, E. (1977) *Phenomenological Psychology*, trans. J. Scanlon, The Hague: Nijhoff.

Husserl, E. (1981) 'Renewal: its problem and method', in *Husserl. Shorter Works* ed. P. McCormick and F. Elliston, Notre Dame: University of Notre Dame Press.

Husserl, E. (1983) *Ideas Pertaining to a Pure Phenomenology and to a Phenomenological Philosophy*, *First Book*, trans. F. Kersten, The Hague: Nijhoff.

Husserl, E. (1989) *Ideas Pertaining to a Pure Phenomenology and a Phenomenological Philosophy*, *Second Book*, trans. R. Rojcewicz and A. Schuwer, Dordrecht: Kluwer.

Husserl, E. (1993) *Cartesian Mediations*, trans. D. Cairns, Dordrecht: Kluwer.

Husserl, E. (1994a) 'Personal Notes,' in *Early Writings in the Philosophy of Logic and Mathematics*, *Collected Works V*, trans. D. Willard, Dordrecht: Kluwer.

Husserl, E. (1994b) *Early Writings in the Philosophy of Logic and Mathematics*, *Collected Works V*, trans. D. Willard, Dordrecht: Kluwer.

Koyré, A, (1978) *Galilean Studies*, trans. J. Mepham, Atlantic Highlands, N.Y.: Humanities Press.

Kusch, M. (1995) *Psychologism*, London: Routledge.

Mahnke, D. (1977) 'From Hilbert to Husserl: first introduction to phenomenology', trans. D. Boyer, in *Studies in the History and Philosophy of Science* 8: 71–84.

Merleau-Ponty, M. (1964) *The Primacy of Perception*, ed. J. M. Edie, Evanston: Northwestern University Press.

Putnam, H. (1987) *The Many Faces of Realism,* La Salle: Open Court.

Putnam, H. (1990) 'After empiricism', in *Realism with a Human Face*, Cambridge, Mass.: Harvard University Press.

Rollinger, R. D. (1996) *Husserl's Position in the School of Brentano*, Utrecht: Dept. of Philosophy.

Rorty, R. (1980) *Philosophy and the Mirror of Nature*, Oxford: Blackwell.

Smith, B. and Woodruff Smith, D. (eds) (1995) *The Cambridge Companion to Husserl*, Cambridge: Cambridge University Press.

Sokal, A. and Bricmont, J. (1997) *Impostures intellectuelles*, Paris: Editions Odile Jacob.

Willard, D. (1979) 'Husserl's critique of extensionalist logic; a logic that does not understand itself,' *Idealistic Studies* IX:2 (May): 143–64.

9 Some scientism, some historicism, some critics

Hayek's and Popper's critiques revisited

Thomas E. Uebel

Arguments against scientism of some form or another shadow the history of the philosophy of social science. Typically such arguments link highly theoretical, indeed meta-theoretical considerations, to practical consequences of a politically and morally odious or salutary nature. They are methodological arguments with a moral-political sub-text. Sometimes, however, such arguments carry an additional sub-text of another type. When the positions criticised are only abstractly characterised and attributed in the main to long dead theorists it is natural to ask who is really meant to be refuted. When over time certain key-phrases lose the resonance that would have helped contemporary readers to identify the addressee of the critique, that sub-text is easily lost. Without the proper contextual understanding of the argument, however, it can become difficult to judge its actual successes and failures.

In this paper I will consider one classical argument against scientism as an example. The argument in question is Friedrich August Hayek's in 'Scientism and the study of society' (1942–44 [1979]), with which I will compare Karl Popper's 'Poverty of historicism' (1944–45 [1961]). I will show that Popper affected a significant reorientation of Hayek's charge, but also that despite the resultant differences in their positive positions there obtained a remarkable agreement between them.[1] Then I will ask and answer the question against which theorists of social science – as opposed to shadowy totalitarian propagandists – Hayek's and Popper's arguments may have been directed. After pointing to an analogy between their critique of their obliquely intended opponents and that launched by critics from the other end of the political spectrum, I will assess the approriateness of Hayek's and Popper's arguments against their 'hidden' opposition.

1

Hayek's 'Scientism and the study of society' was originally published in instalments in *Economica* from 1942 to 1944 and reprinted as Part One of

The Counter-Revolution of Science in 1952. It attacks 'scientism' as the 'slavish imitation of the method and language of science' in social science and argues against 'the mechanical and uncritical application of habits of thought to fields different from those in which they have been formed', namely, of the methods of natural science in the field of social and especially economic phenomena (1942–4 [1952 [1979: 24]]).

Hayek attacked this imitation in three guises. First, he attacked it as 'objectivism', the demand that the social sciences must 'do without the knowledge derived from introspection' (ibid.: 78). Second, he attacked it as 'collectivism', the assumption of 'wholes' 'as definitely given objects about which we can discover laws by observing their behaviour as wholes' (ibid.: 93). Third, he attacked it as 'historicism', the view that history is the only proper science of social phenomena (ibid.: 112) and 'that with the discovery of laws of development [one] had the only key to true historical understanding' (1951 [1952 [1979: 383]]) . More generally, we must add, Hayek rejected empiricism in social scientific concept formation. Thus what Hayek inveighed against as objectivism was reductive, behaviourist naturalism as method, what he inveighed against as collectivism was holism in its ontological and epistemological guises, and what he objected to as historicism was the demand that social science predict large-scale social developments and aid the conscious reconstruction of society. (It is not altogether clear whether these are independent manifestations of scientism or only constitute it jointly. It would seem that Hayek tended towards the former but saw them as intimately connected.)

Against these aberrations Hayek set a version of what we nowadays know as methodological individualism. He decreed that social science must be 'concerned with man's conscious or reflected action, actions where a person can be said to choose between various courses open to him' (1942–4 [1952 [1979: 43]]). This inquiry is restricted to using the agents' conceptualisation of their actions and regards action terminology as irreducible; its task however is to explain only the 'undesigned results of individual action' (ibid.: 69) and to 'constitute' the phenomenal social wholes from individual elements (ibid.: 98). Nonetheless, it provides not merely historically specific theories but universal knowledge (ibid.: 132), since its individualistic elements themselves are held to be universal. The ultimate elements of social structure are 'the structure of men's minds, the common principle on which they classify external events'; social events must ultimately be explained by these elements, which 'are known to us from the workings of our own minds' (ibid.: 59). These individual-psychological elements are foundational for social science in this sense: social science does not have to explicate their explanatory force.

Hayek presented his conception of social science, which he saw most consistently employed in the marginal utility theories of economics, as a descendant of Carl Menger's 'compositive' method and stressed that since the social sciences find their elements directly given, no prior 'resolutive' analysis is required, unlike the natural sciences which must first employ the latter to arrive at its fundamental elements. Hayek left somewhat underdescribed whether this kind of knowledge was straightforwardly empirical first-person psychological or whether this knowledge was synthetic but *a priori*.[2] Presumably Hayek wished to distinguish between knowledge of the structure and content of mental acts of individuals: knowledge of specific contents was empirical whereas that of typical attitudes and their interrelations was a priori. For the constitution of social phenomena both types of knowledge are required. Social institutions are formed by the conservation of certain 'attitudes of the individuals toward each other (or their similar or different attitudes towards physical objects)': their preservation is 'sufficient to preserve a constant structural element' of the social world (ibid.: 59).

Importantly, anti-scientism is *anti-naturalistic*. Against naturalism, Hayek defended the *separatism* of social science with reference to its essential 'subjectivism'. The laws of social science are the laws of the composition of individual actions in so far as their results are unintended. (Actions whose results were intended presumably fall under psychology, whether they are individual or social.) No other theoretical laws are there to be found in social science. In particular, there are no irreducible laws concerning irreducible social wholes; any external regularities that may be established about social phenomena would be not laws, but mere historical facts. That Hayek thus classed Roscher and Schmoller as 'scientistic' in virtue of their 'historicism' (ibid.: 114) brings out nicely his debt to Menger's defence of theoretical economics against the Historical School. For there is also in play Menger's conception of strict theoretical laws, which, *pace* the Historical School, allow of no exceptions.

One might ask, of course, whether what Hayek described as scientism, namely, social-scientific naturalism, must be objectivist, collectivist and historicist in his sense. Clearly, the answer would nowadays be 'no', and we may also wonder whether it was wholly true even in his day. But more still is at issue than questions of methodology. For Hayek, fully blown scientism not only represented an 'abuse of reason' (subtitle of the volume carrying the reprint) that was lamentable for its own sake, but an abuse that was sadly consequential: 'the program of socialism actually derives from this kind of scientistic philosophy' (ibid.: 180). Here we could ask whether socialism requires scientism. Hayek himself noted that the connection between scientism and socialism is not a logical one. Nevertheless he deemed this association important enough to indict scientism for helping to father socialist creed. '[S]cientistic philosophy . . . through its popularisers has done more to create the present trend toward socialism than all the conflicts between economic interests which, though they raise a problem, do not necessarily indicate a particular solution' (ibid.: 179–80). The solution that was anathema to Hayek, of course, was social planning, in particular the proposal to abolish the free market and institute centralist planning. Against this he had what he claimed was a conclusive argument: '[A]ll socialists from Saint-Simon to Marx and Lenin' made the mistake to think that 'a complete concentration of all relevant knowledge' required for such planning 'is possible' (ibid.: 179). Though Hayek did not name it so, his enemy was scientific socialism – the view that a scientifically guided socialist transformation of capitalist economies and class-structured societies is possible – and he quoted Bebel's *Woman in Socialism* to this effect (ibid.: 180).

Here emerges the deep agenda of Hayek's polemic. His concern lay with the normative political debate about the principles along which society should be organised. Hayek sought to show that his political opponents are operating with a deeply flawed conception of scientific knowledge: the topic of social science methodology was but means, not end.

II

Let us now compare Popper's 'Poverty of historicism', originally published also in *Economica,* albeit in instalments from 1944 to 1945, and reprinted in book form in 1957. Popper castigated 'historicism' as 'the approach to the social sciences which assumes that historical prediction is their principle aim, and which assumes that this aim is attainable by discovering the 'rhythms' or the 'patterns', the 'laws' or the 'trends' that underlie the evolution of history.' (1944–5 [1961: 3]) In a sense to be qualified presently, Popper's 'historicism' is Hayek's 'historicism' writ large; for Popper, Hayek's 'scientism' was one version of his own 'historicism', namely the 'naturalist' as opposed to the 'anti-naturalist' one. For both, the attitude of scientism/historicism underwrote totalitarianism by providing its theoretical legitimation and epistemological foundation (Hayek 1951 [1952 [1979: 399]]; Popper 1944–5 [1961: 159]).

Like Hayek before him, Popper sought to nail his opposition on the cross of holism, that is, anti-individualism. Anti-naturalist historicists are mistaken in thinking there to be essences of social wholes that could in some way be intuited like the properties of a *Gestalt*. Anti-naturalistic historicists rely either on an anti-experimentalism that is ungrounded (there obtains no principled difference from natural science in this respect) or on the particularisation of generalisations to certain periods or ages, a view that shows, so Popper wrote, a misunderstanding of the task of theory (to discover strict universal laws). Naturalistic historicists are characterised by their view that the only true social science is history (1944–5 [1961:39]) and that social science is 'the study of the operative forces and, above all, of the laws of social development' (ibid.: 45), with supposedly objective teleologies grounding the normative claim of such studies to guide action (ibid.: 50). Later they were shown mistaken in thinking that there exist laws involving social wholes that are involved in strict laws: so-called laws of development are but statements of trends and these designate a singular fact at best. Popper held that historicism (like scientism) sought laws in the wrong places: there are no historical laws of the development of collectives, but only laws of the aggregation of individual actions. What laws there are to be discovered must be universal with variables ranging only over individuals or states thereof.

Like Hayek as well, Popper was an individualist not only in an ontological and epistemological but also in moral and political sense. Both defended liberal economic freedoms (the right to own means of production) as a precondition of liberty and/or progress. Thus we may note that, in retrospect, both can be seen as critical of the attempt to fashion economics as the axiomatic physics of the social world.[3] Nineteenth century mechanics simply did not translate into a psycho-social equivalent in equilibrium economics. Not only did Hayek and Popper (each for different reasons) reject the epistemological presuppositions of the nineteenth century mechanists (Hayek being apriorist, Popper anti-inductivist), but the delineation of social mechanisms faced a sharp limit which, for Hayek, spelt the impossibility of comprehensive social and economic planning. With individuals as atoms, as it were, the market represents a device that automatically performs a calculation of the values of all goods and services, but this calculation of market values in their totality is impossible for humans to perform. Nevertheless, the market performs an indispensible signalling role in allowing individuals to rationally conduct their affairs in the light of their limited and fallible knowledge of market values. Thus the market serves to coordinate individual actions.[4] The laws of the market provide a social mechanics without Laplacean totalities: the science of unintended consequences of intentional action by necessity had to remain a piecemeal theory of society.

Yet Popper's argument differs in kind from Hayek's, even though there is widespread agreement between them, as we have just seen. Whereas Hayek affirmed the separation of social science, Popper affirmed the 'unity of method' between the natural and social sciences. Even though Popper was concerned to stress his agreement with Hayek at every opportunity (at the time he sought, after all, to leave New Zealand), he could not gloss over this point. He finessed it by invoking Hayek's hero, Carl Menger, in putting forward his distinguishing move (ibid.: 131). Menger's conception of theoretical deductive science is correct for *all* of theoretical science, not just economics or social science. In consequence, of course, the 'fundamental elements' of sociological analysis required to be conceived differently than on Hayek's model: not as psychological data, claiming a foundational place in the explanation of the social, but as explanatory *hypotheses*. For Popper, unlike for Hayek, instrumental rationality and its constitutive attitudes are not given to us synthetic a priori, but are postulated as no less theoretical concepts than those of the fundamental concepts of physics.

Just like the similarities between them, this difference is of considerable importance. Unlike Hayek, Popper was able to start the fight against scientism/historicism from within the naturalist camp, if this is neutrally defined not as scientism but as the denial of the separation thesis (Popper was of course not a naturalist in a stronger sense). Popper's recourse to Menger allowed him to redirect Hayek's attack on naturalism (the anti-separation thesis) and turn it against anti-individualists of all stripes, at the price of dropping Hayek's anti-objectivism. The latter's place was taken by Popper's anti-psychologism (which distinguishes him from Hayek most markedly). Popper's prescriptions for the conduct of social science are different therefore; less contemplative than Hayek, he preferred the term 'piecemeal engineering'.

It is perhaps worth stressing that, if ever Popper tried to minimize his disagreements with another's theory, it was here. Popper deferred wherever possible to Hayek and presented what in fact amounts to a radical reorientation of perspective as something of a clarification of remaining obscurities by way of comments on two extensive quotations from 'Scientism'! All the more emphatic was Popper's agreement with Hayek's animadversion against scientific socialism. Where Hayek argued against the unholy alliance of scientism and socialism, there Popper inveighed against that of historicism and utopianism. Utopianism is closely allied to socialism, of course, as with its replacement of the free market by central planning. Popper followed Hayek in considering this the abandonment of the only instrument which affords rational calculation of courses of action. Both Popper and Hayek alleged wholesale holism on their opponents' part and argued that this gross but widespread misunderstanding of the nature of scientific knowledge had fatal consequences in real life.

III

There are several questions before us now. The first concerns the success of Hayek's and Popper's arguments against scientism/historicism and for their preferred social science methodology. (I will call this their 'manifest' agenda.) The second concerns the success of their attempt to destroy the epistemological foundations of the socialist opponents whose views they delineated. (I will call this their 'latent' agenda.)[5]

Let us begin with the manifest agenda and turn to Hayek first. Few if any academic philosophers had an argument with Hayek's rejection of the teleological and holistic social science that would provide legitimation for totalitarian schemes by furnishing appropriate laws of historical development. It was, after all, on this point that Popper established the continuity of his views with Hayek's and that even sharp critics conceded that Hayek's 'discussion of the logical defects of historicism could hardly be bettered' (Nagel 1952 [1991: 95]). (Whether this critique also effected non-totalitarian socialism, however, was another question, of course, on which no wide agreement has been forthcoming.) But it was Hayek's contention that the separation thesis is correct and its denial deleterious that prompted the most vocal dissent. Thus Ernest Nagel argued in his influential review of *Counterrevolution* that Hayek 'has not established his central contentions – neither his thesis that the natural and the social sciences employ radically different methods, nor his claim ... that socialism and authoritarianism are logically related to scientism' (ibid.: 94–5).[6] In other words, Nagel gave a sharply differentiated evaluation of the two parts of Hayek's manifest agenda.

Nagel rejected Hayek's argument for subjectivism for it 'confound[s] the genesis for our ideas with their validity' when Hayek 'supposes that because the alleged ways of hitting upon the explanatory data in the natural and social sciences are different, the canons for validating assumptions concerning the existence and operation of those data must also be different' (ibid.). Hayek's argument against collectivism was faulted for its disregard of examples from the natural sciences where 'complex "wholes" . . . are the objects of study' and for its failure to appreciate the theoretical nature of 'human attitudes (as distinct from their specific manifestations)' (ibid.: 93). Nagel concluded that 'what Mr Hayek calls the analytic and the compositive methods are employed in both the natural and the social sciences, so that neither method can be taken as distinctive of either branch of inquiry' (ibid.: 94). Hayek's contention that the importation of the methods of natural science into social science leads to historicism thus stands refuted. In a similar fashion, Nagel contended, Hayek failed to show that non-teleological social science was 'incapable of accounting for coherent and valuable social arrangements except as products of overt planning' (ibid.). In sum, Hayek had 'certainly not made good his claim that the extension of the methods of the natural sciences into social inquiry is an abuse of reason' (ibid.: 95)

Comparing Nagel's and Popper's reactions to Hayek's critique of scientism itself, it is not immediately clear what Nagel added to Popper's earlier reorientation of Hayek's arguments, apart from changing the rhetoric from that of basic agreement with a minor dissent thrown into the bargain to that of major dissent with a minor agreement. While Nagel defended theorising about social wholes, he stressed that the legitimate examples of complex wholes in natural science are conceived as 'theoretical "constructs" out of more familiar observable components'. (This hardly amounts to a disagreement with the methodological individualism that Popper championed.) Nagel's position concerning Hayek's anti-objectivism also broadly agreed with Popper's both in terms of epistemological strategy and methodological prescription: Hayek's aprioricism is roundly rejected and the hypothetical character of theoretical social science, to be tested by predictions, is reaffirmed by both. On the first question then they agreed: Hayek's simultaneous argument for subjectivism and the separation thesis and against scientism, understood as affirming as the unity of scientific method, failed. Why then the much sharper dissent on Nagel's part?

One wonders whether Popper muted his criticism at least in part because he shared Hayek's latent agenda. Once they were stripped of their anti-objectivism and suitably recast, Hayek's arguments against collectivism and historicism were taken by Popper to constitute a conclusive argument which demolished the epistemological basis of totalitarian social theory. Popper thus retained the line of argument linking epistemologies and social theories, which Nagel rejected. For Popper, historicists were forced into utopianism due to their holism; what he failed to ask was whether all socialists had to be historicist, utopian and holist in just the sense alleged. Nagel at least left this issue open.

Yet how did Popper himself do on the manifest agenda? He clearly shared Hayek's laurels in combatting any teleological and holistic social science that would legitimise totalitarianism by suitable laws of historical development. Since he also did not attack naturalism, the negative part of his manifest agenda was wholly successful. I will turn to its positive aspect after considering the success of his and Hayek's latent agenda.

IV

The question arises, as I already noted, whether the positions of Popper's historicist and Hayek's scientistic social scientist mark out a kind of viewpoint held by thinkers not in the sway or pay of totalitarianisms of the right or the left. But we may also ask just what kind of 'practical' social science really is ruled out by Hayek's and Popper's principled arguments. Equal unclarity rules here, so let us briefly pursue this question first.

Nagel held social science to be capable of 'accounting for coherent and valuable social arrangements', as we saw. Whether this constitutes a sharper disagreement with Hayek's anti-interventionism than Popper's advocacy of piecemeal engineering is not clear; both certainly seem *prima facie* compatible. In fact, it is nor even entirely clear whether Popper's own conception, as Popper himself claimed (1944–5 [1961: 61]), or Nagel's reference to 'theories of planned economy . . . within the framework of marginal utility analysis' (Nagel 1952 [1991: 94]) constitute a radical break with Hayek's critique. As J. W. N. Watkin's very favourable review – which discounted his criticism of Hayek's subjectivist methodology as minor – pointed out, Hayek also wrote that '[a]s we learn to understand the spontaneous forces, we may hope to . . . modify their operations by proper adjustment', stressing that 'there is all the difference between thus utilising and influencing spontaneous processes and an attempt to replace them . . . by conscious control' (Watkins 1953 [1991: 106]). Similarly, Hayek applauded the theorists obliquely referred to by Nagel – Lange and Taylor's and Dickinson's models of 'competitive' or market socialism – as 'courageous attempts to remold socialist doctrine' (1940 [1948: 208]), thus seemingly placing them out of reach of his anti-scientistic argument (but still rejecting them on other grounds).

It would seem then that the positions correctly attacked by Hayek and Popper are so extreme as to count as virtual straw-men for all but the crudest of anti-communist crusades. Just in the context of these, of course, Hayek's and Popper's polemics were extremely successful: Hayek and Popper became Cold War figureheads for liberty and democracy. Hayek was adopted by American libertarians who made him into still more of an anti-interventionist than he already was, but several decades later he was recognised also by European conservatives who had begun to feel the loss of the authority of tradition. Popper, of course, was even more widely championed for celebrating the 'open society', his influence reaching even into social democratic circles.

One might remark that academic economics was hardly in danger of capitulating to socialist theorising, but this would be to overlook not only Keynesianism but also that by the end of 1945 the Labour Party had been elected to power in Britain, with Beveridge's concept of 'planning for full employment' and the nationalisation of basic industries moving ahead. Thus in his Finlay lecture of 17 December 1945, Hayek complained that 'we are in fact rapidly moving from a society of free individuals toward one of a completely collectivist character' (1946 [1948: 1]). From his perspective, the battle only intensified during the forties and more than justified the publication of 'Scientism' with related articles as *Counterrevolution* in 1952 (after he had moved to Chicago). By contrast,

Popper's own republication of 'Poverty' in book form in 1957 was most likely at least partly motivated by his desire that the justified criticisms made by Hayek should not swamp all predictive social science and piecemeal social engineering.[7]

But to return to the main theme of their latent agenda: how successful were Hayek and Popper in their aim of pulling the methodological rug from under the socialist opposition? Here we must note that it is doubtful whether Labour's captains of the mixed economy would have thought themselves to be under attack by Popper's assault on utopianism. Precisely that it was a mixed economy they championed protected them from Popper's charge. As for Hayek's charge, derived from Mises, that 'without the price system we could not preserve a society based on such extensive division of labor as ours' (1945 [1948: 89]), again even old Labour was untroubled since they did not challenge it. So there is a clear incongruity between the weapons trained by Hayek and Popper on the socialist enemy and British socialism of the post-war period. It would, of course, have been hard to call the latter totalitarian, so at least Popper may not have wished to have his criticism apply to it. This raises a curious question. Who were the contemporary theorists of scientific socialism that Hayek and Popper gunned for? Their explicitly methodological rhetoric suggests that it was *not only* opposition to party-hacks and other shadowy advocates of dialectical materialism that inspired them. Thus we must ask who of the proponents of socialism that Hayek and Popper deemed to be appropriate objects of their critique were *also* theorists of social science?

Let us first note which more or less contemporary socialist theorists are referred to by Hayek. He mentions E. F. M. Durbin, L. T. Hobhouse, Joseph Needham, Karl Mannheim, Bertrand Russell and Otto Neurath (1942–4 [1952 [1979: 119n, 155–6, 179, 170n]]). Are they all representatives of centralist economic planning that profess scientism? Amongst those mentioned only Russell and Neurath fit the bill. But whereas Hayek could only pin some broadbrush pro-planning remarks on Russell, Neurath once had been famous (if not notorious) for his proposals for administrative economies made during the German revolution of 1918–19.[8] Thus Hayek remarked in the context of his polemic against social planning: 'The most persistent advocate of such *in natura* calculation is, significantly, Dr Otto Neurath, the protagonist of modern "physicalism" and "objectivism"' (ibid.: 170n; cf. 78). We do well to remember here what Hayek does not note explicitly, that Neurath was a central villain of Hayek's in his introduction to his edition of essays *Collectivist Planning* (1935) and even still before that of von Mises in his original paper on the calculation problem of 1920 and his book-length refutation of socialism built thereon (1923). To popularise these 'results' of the continental calculation debate in Britain was in fact the point of Hayek's *Collectivist Planning*, with his own contributions to it (and a follow-up) reprinted in *Individualism and Economic Order* (1948). Neurath apparently is the one twentieth-century theorist upon whom Hayek can with at least *prima facie* plausibility pin the joint sins of scientism and comprehensive planning theory.

As in the case of Hayek, we may ask who the utopian theorists are that Popper condemned to irrationality. They too remain a strangely shadowy bunch. Mannheim is convicted by association, but other thinkers behind the Stalinist *Gestalt* Popper outlined are not identified. Neurath is only once referred to – most curiously, in the chapter dismissing anti-naturalist historicism! – for his supposed confusion about the scientific method (ibid.: 103), but that he may neverthless be one of the intended objects of Popper's polemic as well only becomes clear once we note Popper's deferential reference to Hayek's *Collectivist Planning* in support of his own argument for the inherent irrationality of utopian, holistic planning. Neurath is one twentieth-century theorist upon whom Popper can foist, with at least *prima facie* plausibility, the joint acceptance of historicism and utopianism.

Would the conclusion that, besides unspecified dialectical materialists, it is none other than Otto Neurath whom Hayek and Popper are arguing against, amount to more than a historical curiosity? It is curious, but not just that. As it happens, for both Hayek and Popper, Neurath was a central figure of opposition long before they took issue with his overall theory of social science. Both of their *Economica* pieces represented an expansion of earlier, differently located criticisms of his proposals for administrative economies (Hayek 1935) and his conventionalist-pragmatist conception of the empirical basis of science (Popper 1934). But Hayek's and Popper's charges of scientism and historicism, though obliquely *ad hominem*, are not really personal. It is the philosophy of social science propounded by logical empiricism that was placed in the dock. To be noted here is that at the time when Hayek's and Popper's critiques were written and first published Neurath was more or less the sole representative of logical empiricist theory of social science (before, that is, the latter was sanitised by Richard Rudner in the 1950s and 1960s).[9] If this is correct we must refocus our previous question. Did Hayek's and Popper's arguments work against Neurath's logical empiricist philosophy of social science?

First let us note, however, that Hayek's own memories of the Vienna Circle and Neurath confirm our conclusion. Hayek claims to have been informed about the goings-on in the Circle by a member of his own *Geistkreis* who was at the same time a member of the Vienna Circle, Felix Kaufmann. Since despite corrrespondence with Neurath on these matters Kaufmann persisted in misportraying 'physicalist' social science in his

(1936), we may doubt, however, whether the intelligence Hayek received got the goings-on in the Circle just right.[10] Certainly his account of Neurath's theory of social science appears strongly influenced by Kaufmann's account. In any case, Hayek consciously kept his distance from the Vienna Circle.

> What dissuaded me is that the social scientists, the science specialists in the tradition of Otto Neurath . . . were so extreme and so naive on economics; it was actually through them that I became aware that positivism was just misleading in the social sciences. I owe it to Neurath's extreme position that I recognized it wouldn't do. And it took me a long time, really, to emacipate myself from it. It was only after I had left Vienna, in London [Hayek took up his position at the LSE in 1931], that I began to think systematically on problems of methodology in the social sciences, and I began to recognize that positivism in that field was definitely misleading.
>
> (Hayek 1994: 50)

Popper, for his part, as is well known, was forever engaged in differentiating himself from the logical empiricists. Since, in addition, he still owed a response to Neurath's critique of his *Logik der Forschung*, 'Pseudo-rationalism of Falsification' (Neurath 1935b), a suitable treatment of Neurath's theory of social science must have been an inviting prospect. In the case of both Hayek and Popper then it may very well have been – no conspiracy story this! – the very centrality of Neurath to their methodological undertakings that brought about, by way of overcompensation, his marginalisation in the way their arguments were developed. They placed *die Sache selbst* – the 'fascinating intellectual structure', as Popper put it (1957 [1961: vii]), of historicism/scientism – into centre stage.

In any case, the rhetoric of disembodiment and decontextualisation worked its ruse brilliantly: anybody tarred by the brush of such principled criticism as theirs certainly had their work cut out for them. Neurath, of course, hardly had time to register his disagreement with 'Scientism' – he died in December 1945 – and to seek, repeatedly but unsuccessfully, to arrange for a public debate of the issues with Hayek.[11] As far as Popper's polemic was concerned, of course, Neurath long knew to expect mischief from him and concentrated on what at the time seemed the more important detractor.[12] Of course, when Hayek's and Popper's polemics were republished in book form, Neurath had already faded from the consciousness of the wider public and was remembered only vaguely by the philosophical community at large, more likely than not as the 'wild man' of logical positivism. Would it not have been somewhat indecorous of Hayek and Popper to have 'outed' their old opposition more explicitly then? Their charge, in any case, remained unanswered.

V

Now certainly it is a noteworthy distinction to have been the true whipping-boy of two of the most distinguished polemics in twentieth-century philosophy of social science. It all depends, of course, on what we make of these polemics. If they were conclusive, the honour accorded would be onerous. Before assessing them, let me note that different but structurally analogous arguments were raised also in quite different quarters.

Under the charge of 'scientism', we saw, the philosophy of science of the Vienna Circle was accused by Hayek of 'abuse of reason'. More was at issue in Hayek's charge than mere disagreement with the philosophy of social science. His complaint ultimately concerned the politics of those whom he deemed scientistic thinkers. We also saw that what Hayek termed 'scientism' is, despite their differences, closely related to what Popper termed 'historicism' and which Popper identified as conceptual midwife to totalitarianism. Both behind Hayek's technical-philosophical disagreement with so-called 'scientism' and Popper's with 'historicism' stood a clearly political animadversion.

But such double-barrelled criticism was levelled at the Vienna Circle not only from the Hayek–Popper camp, but also thinkers from the other end of the political spectrum. Notable here is not only Lenin's notorious attack on empirio-criticism (Lenin 1908), which, by singling out, amongst others, the young Philipp Frank, provided the *locus classicus* for later Marxist derogations of logical empiricism. Thus already condemned as 'Machists' by Lenin, Nicholai Bucharin's address to the Second International Congress of History of Science and Technology in London in 1931 finished the job. He accused the Viennese neopositivists of reducing cognition to the production of tautologies, of reducing the human subject to a merely passive observer and of denying, by virtues of their fancy for 'epistemological Robinson Crusoes', the essentially social nature of man; ultimately their sin was the refusal to see that the unity of scientific method consisted not in the rigorous application of the so-called 'new logic' but in that of dialectical materialism (Bucharin 1931: 12, 17, 21, 32; cf. 1934 [1935: 27]). These were of course the criticisms that were regularly reproduced in the USSR, and later the GDR and other Warsaw block countries, still decades after their author himself had fallen from favour and victim to Stalin's purges.

Yet also closer still to home, the Vienna Circle came in for rough treatment. Max Horkheimer, director of the Frankfurt Institute for Social

Research, who had been approached by Neurath with a view to collaboration in their shared exile from Nazi Germany, also launched a savage critique in his institute's house journal in 1937, a critique that apparently is still taken as definitive by contemporary Critical Theorists like Jürgen Habermas whose formative academic years were at least partly marked by his participation in the German *Positivismusstreit* (Adorno *et al.* 1969). That Horkheimer repeatedly refused to print Neurath's response to his attack – it remains unpublished to this day – may well have helped it gain currency, for its characterisation of Vienna Circle philosophy has little to recommend it in terms of accuracy. [13]

As with Hayek, the ultimate aims were the socio-political consequences that were thought to follow from the Circle's 'scientific world-conception'. Unlike Hayek, of course, and rather in line with its Soviet critics, Horkheimer criticised what he deemed its bourgeois quietism. Thus he remarked, apropos of Hahn's claim that scientific thought cannot transcend the realm of experience:

> This principle is particularly significant in a world whose magnificant exterior radiates complete unity and order while panic and distress prevail beneath. Autocrats, cruel colonial governors, and sadistic prison wardens have always wished for visitors with this positivistic mentality. If science as a whole follows the lead of empiricism and the intellect renounces its insistent and confident probing of the tangled brush of observations in order to unearth more about the world than even our well-meaning daily press, it will be participating passively in the maintenance of universal injustice.
>
> (Horkheimer 1937 [1972: 151])

And after exposing 'the primitive misconceptions exhibited by the modern empiricists', Horkheimer called theirs:

> a philosophy that resolves not to make any essential distinction between the conspiracy of brutal despots against all human aspiration to happiness and freedom, on the one hand, and the struggles to defeat these tyrants on the other; a philosophy that reduces the two to the abstract concept of the 'given' and even glorifies such conduct as objectivity.
>
> (Horkheimer 1937 [1972: 177–8)

Clearly Horkheimer did not pull his punches. In his own way, he accused the Vienna Circle of the abuse of reason, namely of 'confounding . . . calculatory with rational thinking' (but unlike Hayek and like the Soviet critics advising of the need for dialectical thought) and he concluded that 'science becomes naively metaphysical when it takes itself to be the knowledge and the theory and even goes so far as to disparage philosophy, that is, every critical attitude towards science' (ibid.: 183).

Whereas Hayek and Popper turned logical empiricists like Neurath into more or less willing stooges of Stalinism, Bucharin and Horkheimer turned them into unwitting helpmates of fascism and National Socialism. Their abuse of reason, they all complained, promoted what they viewed as the abuse of humanity. This remarkable agreement between these critics of Vienna Circle philosophy, whose views on their shared central concern – the normative principles of social organisation – were diametrically opposed, should make us suspicious. Not only was the Circle's epistemological and metatheoretical position rejected by the critics from both the hard left and the staunch right, but both did so to further their own views on a quite different topic: precisely the norms of social organisation.

Certainly both sets of critics cannot be right simultaneously. It is hard to see how one could be conservatively quietist and politically conformist and at the same time be part of the vanguard of the class struggle. Of course, there *were* clear disagreements between the Vienna Circle and their opponents about the role of philosophy. They did not agree with Horkheimer that 'science and its interpretation are two different things' (ibid.: 183–4). Similarly, they denied the separate standing of the *Geisteswissenschaften* that Hayek defended and they rejected Popper's attempt to rehabilitate metaphysics as meaningful and deductive closure as the true mark of genuine knowledge. Arguably, however, these differences did not arise from misconceptions on part of the logical positivists.

Of course, the two sets of critics might also both be wrong. That in effect was the position of Nagel, but notably so, he did not set out to defend Neurathian philosophy of social science as such. In any case, two arguments are needed for making the case that both sets of critics are mistaken. The first argument must attempt to meet head-on the central challenge of the left critics that the scientific world-conception absolutizes natural scientific knowledge and denigrates the normative realm of ethical values, justice and the class-struggle (or whatever is deemed to take its place). The second argument must show that, in a word, politicised logical empiricist philosophy of social science does not fall foul of reasonable strictures on objective social science. Elsewhere I assembled the materials for the first argument; here I seek to advance the second argument to the degree required by Hayek's and Popper's challenges.[14]

VI

Are Hayek's and Popper's charges against Neurath correct? Neurath self-consciously confessed to 'scientism' (Neurath 1935a [1983, 115]) and 'scientific utopianism' (Neurath 1919; 1944, § 13). But is it his scientism that is attacked by Hayek and his utopianism that is attacked by Popper?

For present purposes we can put the thorny issue of Neurath's proposals for administrative economies and his 'rational economic theory' (Neurath 1935c) to one side and concentrate on whether Neurath's social scientific methodology does fall prey to Hayek's and Popper's arguments. If, as I shall argue, it does not, then Hayek's and Popper's critiques misfire, whatever the nature of Neurath's contentious economics. Note also that the question is of interest not only for students of Neurath, or, since he was their social science specialist, of the wider left-wing of the Vienna Circle. The question whether Neurath's social scientific methodology is really refuted is also of systematic interest in that a negative answer would provide an alternative conception of social science to that of Hayek or Popper, who true to their Mengerian roots can only conceive of one true social science. In other words, Neurath would show us by example both the price that one may have to pay for dissent from individualism as conceived by Hayek and Popper, and what could be attained for that price.

As we saw, already Popper and Nagel found fault with Hayek's argument for subjectivism and against objectivism. But did Hayek succeed against Neurath? There are a number of miscomprehensions we can point to. Hayek wholly misunderstood the point of Neurath's 'physicalism' and his 'social behaviourism' when he read it as a reductivist doctrine that forbade all references to mental states. All it did forbid was reference to disembodied mental states and contents such that attributions would become in principle untestable. Thus Neurath wrote:

> While avoiding metaphysical trappings it is in principle possible for physicalism to predict future human action to some degree from what people 'plan' and 'intend' ('say to themselves'). But the practice of individual and social behaviourism shows that one reaches far better predictions if one does not rely too heavily on these elements, which stem from 'self-observation', but on others which we have observed in abundance by different means.
>
> (Neurath 1936a [1981: 714]).

Neurath's naturalistic methodology did not preclude reference to intentional phenomena. Since Hayek pinned Neurath's scientism explicitly only on his physicalism, his argument against him does not hold up any better than against a naturalist like Nagel. But for the record we may note also that the charges of collectivism and historicism do not apply to Neurath.

Neurathian social science also does not condone reference to the irreducible collective entities like '*Volk*' or '*Volksgeist*': indeed, to combat such talk within the academia and even more outside of it may be said to constitute one of the most central motive forces of his social scientific naturalism. It was just by allowing such talk that the contemporary proponents of separatist *Geisteswissenschaft* were regarded by the Vienna Circle as providing metaphysical pseudo-foundations for fascism and Nazism. This is a major point of agreement with Hayek's and Popper's later arguments which the latter two never mentioned. (Compare also Neurath's opposition Plato's social theory which was applauded, incidentally, by Russell, who did not shy from ridiculing other theses of Neurath's.)[15] Similarly, it may be noted that the later Neurath, albeit in his correspondence, even criticised fellow logical empiricists like Hempel and Zilsel for their seeming attempts to delineate strict laws of history. Moreover, the later Neurath in particular was concerned – somewhat to the chagrin of his colleages – to stress the principled limits to predictability in the social sciences, which he saw as arising from the phenomenon of reflexive predictions, a phenomenon he commented upon already before the First World War.[16] If that is 'scientism', we may ask, where are *its* critics?[16]

So how does Neurath's conception stand up to Popper's redirected criticism? For the reasons just adduced, he also escapes his charge of historicism in the strict sense. Yet as noted, Popper deemed him caught in the wider net of the arguments he spun. He convicted him of mistaking the nature of theoretical science by allowing social science to aim for generalisations that only hold 'within "the present cosmological period"' (Popper 1944–5 [1961: 103]). The quotation is not exact, but Popper's attached footnote correctly refers to Neurath's paper at the Second Congress for Unified Science where a 'relativisation and historicisation' of social scientific generalisations is suggested (Neurath 1936b [1981: 772]). For Popper, Neurath counted as a historicist because he still hankered after what might be called 'soft' historical laws. But are such relativised laws laws of history at all? At least a different reading of the Neurathian enterprise is possible, but it is not one Popper was inclined to pursue. He wrote:

> It would not be a sign of laudable scientific caution if we were to add such a [limiting] condition, but a sign that we do not understand scientific procedure. [Here follows the footnote reference to '[h]istoricists' like Neurath.] For it is an important postulate of scientific method that we should search for laws with an unlimited realm of validity. If we were to admit laws that are themselves subject to change, change could

> never be explained by laws. It would be the admission that change is simply miraculous. And it would be the end of scientific progress; for if unexpected observations were made, there would be no need to revise our theories: the *ad hoc* hypothesis that the laws have changed would 'explain' everything.
>
> (Popper 1944–5 [1961, 103])

Clearly, that it is 'an important' task to find exceptionless laws does not mean that it is the only task for science, unless it is Mengerian theoretical science that we have in mind. But why should the latter be accepted without argument?

To unhinge Hayek's and Popper's argument that social science must proceed precisely as they prescribe, one must challenge the presupposition on which their prescription proceeds. Contemporary debate has effected just this by challenging Popper's and Hayek's conception of what scientific laws are worth aiming for (albeit not under this heading). Contemporary theorists argue that universal laws of the Mengerian variety are by no means what the social sciences should aim for exclusively.[17] One issue that arises here is how *ceteris paribus* laws are interpreted: it becomes important that they are not read as poor cousins of universal laws but considered as epistemically valuable in their own right. It is too early still to predict the end of this debate, but note that now we are squarely back to discussing the issues that enlivened the notorious *Methodenstreit* over a hundred years ago between Menger and Schmoller. One of the sharpest distinctions between theoretical and historical economics for Menger was the absolute universality of theoretical laws!

It does not lack a certain irony therefore that not only did already Neurath make this 'contemporary' move against Popper's later prescriptions, but that he made that move being very well acquainted with the classical positions in the *Methodenstreit*. Concerning that debate, Neurath noted early on that in principle, 'there is no reason to think of historical and theoretical research as opposites, it would not even be practical to conceive of each in isolation from the other' (Neurath 1911: 113). Far from betokening a simple misunderstanding of what science is all about, Neurath disagreed with Popper's absolutist conception of scientific knowledge. I stress this point because without their Mengerian presupposition Hayek's and Popper's arguments against Neurath as a historicist collapses. For if it be granted to social science to aim for less than universal laws, then there is no bar to investigating 'mid-range' generalisations concerning social phenomena and *ceteris paribus* laws qualified by temporal indexes. Since such generalisations may well refuse to trade in ontologically irreducible wholes, that is all that is required to break the deadlock that

Hayek's and Popper's impose on the social sciences: be individualist and universalist or unscientific.

VII

The charge of scientism, we have seen, is not single- but multi-valued: just what the charge is depends upon who is making it – and against whom. Its rebuttal accordingly must take different forms. Hayek's argument against scientism, we saw, does not succeed in general; much less, I argued, does it succeed against its obliquely intended opponent, logical empiricist philosophy of social science of the Neurathian variety (even though we may applaud its force against unspecified party-political propagandists). But even as redirected by Popper against historicism, the oblique argument against Neurath's conception of social science fails (while again doing sterling service in the propaganda wars). Neurath's logical empiricist philosophy of social science might still be accused of some other version of scientism, of course, but that would require different arguments. Hayek's and Popper's, in any case, fail and we've even begun to see good reason for Neurath's transgression of the restrictions they imposed.

In closing, let me stress that my argument has not been directed against the possibility of pursuing social science in the way favoured by Hayek and Popper, but their contention that it *has* to be done their way. It was on the neglect of this small but significant difference, that their common rhetoric and their apparent successes rested. In so far as logical empiricist philosophy of socal science was caught up in the melée prompted by their polemics, the loss has been ours. For current debates about, for instance, social scientific separatism or even the legitimacy of engaged social science proceed wholly undisturbed by Neurath's and the logical empiricsts' actual arguments – which are, as we have begun to see, considerably more interesting than their routinely refuted caricatures.[18]

Notes

1 Urbach (1978, 1987) has subjected Popper's strict 'proof' of the thesis of historicism – 'for strictly logical reasons, it is impossible for us to predict the future course of history' (Popper 1957 [1961: v: Preface to reprint]) – to systematic criticism. I will be concerned with a different aspect of Popper's theses. My concern is not with Popper's supposed refutation of historicism in this specific sense (effected, in any case, in different papers) but his attempt (in 'Scientism' itself) 'to show . . . that historicism is a poor method – a method which does not bear any fruit' (ibid.: v). It is Hayek's and Popper's prescriptions for and restrictions on the conduct of social science that concerns me here.

2 A reference to von Mises' *Human Action* and thus his praxeology, which would support the latter reading, was added in the reprint (ibid.: 53n), yet all along

Menger was quoted as specifying the basic elements as empirical: 'human individuals and their intentions' (ibid.: 66n).

3 On the latter see Mirowski 1989.

4 As it happens, it is not clear whether Popper follows Hayek's belief – typical for Austrian economics in the narrow sense and different from the neo-classical synthesis (Hoover 1988) – that the market cannot be assumed to clear, that a general equilibrium cannot be assumed to obtain: piecemeal theorising is prescribed already on general epistemological grounds.

5 Both questions arise both systematically and historically, of course. It may well be that a doctrine is for a period deemed to have dealt adequately with its opposition when either the opposition was only caricatured or the systematic argument is faulty.

6 While Hayek did not claim that socialism and authoritarianism required scientism, as we saw – 'logically, methodological collectivism and political collectivism are distinct' – he did claim that 'the collectivist method . . . leads thus directly to political collectivism' (1942–44[1952(1979:161)]). It is the latter claim on which Nagel somewhat infelicitously focused.

7 If in the early 1940s Popper was concerned to establish agreement with Hayek, in the late fifties he was presumably more concerned to present his own conclusions and methodological prescriptions. Thus the new Preface dates his original argument to before 1936 when, the reader is now told, he read an early version of 'Poverty' to Hayek's seminar during a visit to London (when Popper scouted the possibilities of leaving Austria for Britain). But the body of the text remained unaltered in the main, and with it Popper's original rhetorical strategy of foregrounding their agreement.

8 For descriptions of Neurath's socialisation plans see Part 1 of Cartwright, Cat, Fleck and Uebel 1996; for discussions of his economics see Chaloupek 1990 and O'Neill 1995 and 1998.

9 Karl Menger (the son of Carl) has not been forgotten: given the distance he put between himself and the Vienna Circle, with which he was associated early on, in response to the political tenor of its manifesto of 1929, he hardly counted as a logical positivist or empiricist at the time.

10 See Neurath's correspondence with Kaufmann in *Otto Neurath: Correspondence*, Vienna Circle Archive, Rijksarchief Noord-Holland, Haarlem, Netherlands.

11 See Neurath's correspondence with Hayek in *Otto Neurath: Correspondence*, Vienna Circle Archive, Rijksarchief Noord-Holland, Haarlem, Netherlands.

12 See scattered remarks about Popper in Neurath's correspondence with Carnap, *Rudolf Carnap: Correspondence*, Archives of Scientific Philosophy, Hillman Library, University of Pittsburgh.

13 For the reconstruction of the dispute between Horkheimer and Neurath see Dahms 1994 and comments in Uebel 1994.

14 A systematic treament of Neurath's theory of social science is given in Uebel 1997; the materials for answering Horkheimer's charge are assembled in Uebel 1998.

15 See Neurath and Lauwerys 1945 and Russell 1945.

16 For more on these points, including references, see Uebel 1997.

17 For instance Kincaid 1996.

18 Without wishing to attribute agreement, I should like to thank Greg Ransom for correspondence and Daniel Hausman and Hillel Steiner for comments on an earlier draft.

Bibliography

Adorno, T. W. *et al.* (1969) *Der Positivismusstreit in der deutschen Soziologie,* Neuwied/Berlin: Luchter, trans. as *The Positivist Dispute in German Sociology*, London: Heinemann, 1976.

Bucharin, N. (1931) 'Theory and practice from the standpoint of dialectical materialism', in Delegates of the USSR to the International Congress of the History of Science and Technology, *Science at the Crossroads,* London, 2nd edn. London: Frank Cass, 1971: 11–33.

Bucharin, N (1934) 'Marx's teaching and its historical importance', in N. Bucharin *et al, Marxism and Modern Thought* [in Russian], Moscow, trans. London: Routledge, 1935: 1–90.

Cartwright, N., Cat, J., Fleck, L. and Uebel,T. E. (1996) *Otto Neurath: Philosophy between Science and Politics*, Cambridge: Cambridge University Press.

Chaloupek, G, K. (1990) 'The Austrian Debate on economic calculation in a socialist economy', *History of Political Economy* 22: 659–75.

Dahms, H.-J. (1994) *Positivismusstreit,* Frankfurt a.M.: Suhrkamp.

Hayek, F. A. (ed.) (1935) *Collectivist Economic Planning*, London, Routledge.

Hayek, F. A. (1940) 'Socialist calculation III: the competitive "solution"', *Economica* 7, repr. in Hayek 1948: 181–208.

Hayek, F. A. (1942–44) 'Scientism and the study of society', *Economica* 9–11, repr. as Part One of Hayek 1952 [1979: 17–182].

Hayek, F. A. (1945), 'The use of knowledge in society', *American Economic Review* 35, repr. in Hayek 1948: 77–91.

Hayek, F. A. (1946) *Individualism: True or False,* Oxford, Blackwell, repr. in Hayek 1948: 1–32.

Hayek, F. A. (1948) *Individualism and Economic Order,* Chicago, University of Chicago Press.

Hayek, F. A. (1951) 'The counter-revolution of science', *Measure*, repr. as Part Two of Hayek 1952 [1979: 183–400].

Hayek, F. A. (1952) *The Counter-Revolution of Science. Studies on the Abuse of Reason*,New York: Free Press; 2nd edn, Indianapolis: Liberty Fund, 1979.

Hayek, F. A. (1994) *Hayek on Hayek*, ed. S. Kresge and L. Wenor, London: Routledge.

Hoover, K. D. (1988) *The New Classical Macroeconomics,* Oxford: Blackwell.

Horkheimer, M. (1937) 'Der neueste Angriff auf die Metaphysik', *Zeitschrift für Sozialforschung 6*, trans. as 'The latest attack on metaphysics' in M. Horkheimer, *Critical Theory*, New York: Seabury Press, 1972: 132–87.

Kaufmann, F. (1936) *Methodenlehre der Sozialwissenschaften*, Vienna: Springer.

Kincaid, H. (1996) *Philosophical Foundations of the Social Sciences*, Cambridge: Cambridge University Press.

Lenin, V. I. (1908) *Materialism and empirio-criticism* [in Russian], transl. New York: International Publishers, 1927.

Mirowski, P. (1989) *More Light Than Heat*, Cambridge: Cambridge University Press.

Mises, L. von (1923) *Die Gemeinwirtschaft*, 2nd edn 1932, trans. as *Socialism. An Economic and Sociological Analysis*, London: Cape, 1935, enlarged edn 1951.

Nagel, E. (1952) '[Review of] Hayek, *The Counter-Revolution of Science*', *Journal of Philosophy* 49, repr. in Wood and Woods 1991: 90–5.

Neurath, O. (1911) 'Nationalökonomie und Wertlehre', *Zeitschrift für Volkswirtschaft, Sozialpolitik und Verwaltung* 20: 52–114.

Neurath, O. (1919) 'Die Utopie als gesellschaftstechnische Konstruktion', in *Durch die Kriegswirtschaft zur Naturalwirtschaft*, Munich: Callwey, trans. as 'Utopia as a social engineer's construction' in O. Neurath, *Empiricism and Sociology*, ed. M. Neurath and R. S. Cohen, Dordrecht: Reidel, 1973: 150–5.

Neurath, O. (1935a) 'Einheit der Wissenschaft als Aufgabe', *Erkenntnis* 5, trans. as 'The unity of science as a task' in Neurath 1983: 115–20.

Neurath, O. (1935b) 'Pseudorationalismus der Falsifikation', *Erkenntnis* 5, trans. as 'Pseudorationalism of falsification' in Neurath 1983: 121–31.

Neurath, O. (1935c) *Was bedeutet rationale Wirtschaftsbetrachtung?*, Vienna: Gerold, trans. as 'What is "rational economic theory"?' in B. McGuinness (ed.) *Unified Science,* Dordrecht: Kluwer, 1989: 67–109

Neurath, O. (1936a) 'Mensch und Gesellschaft in der Wissenschaft', *Actes du Congres Internationale de Philosophie Scientifique, Sorbonne, Paris 1935*, Facs. II 'Unite de la Science', Paris: Herman, repr. in Neurath 1981: 711–18.

Neurath, O. (1936b) 'Soziologische Prognosen', *Erkenntnis* 6, repr. in Neurath 1981: 771–6.

Neurath, O. (1944) *Foundations of the Social Sciences,* Chicago: University of Chicago Press.

Neurath, O. (1981) *Gesammelte philosophische und methodologische Schriften*, ed. R. Haller and H. Rutte, Wien: Hölder-Pichler-Tempsky.

Neurath, O. (1983) *Philosophical Papers 1913–1946*, ed. and trans. R. S. Cohen and M. Neurath, Dordrecht: Reidel.

Neurath, O. and Lauwerys, J. A. (1945) 'Plato's *Republic* and German education', *Journal of Education* 77: 57–9, 222–4, 394.

O'Neill, J. (1995) 'In partial praise of a positivist. The work of Otto Neurath', *Radical Philosophy* 74: 29–38.

O'Neill, J. (1998) *The Market. Ethics, Knowledge and Politics*, London: Routledge.

Popper, K. R. (1934) *Die Logik der Forschung*, Wien: Springer, trans. with addenda by author as *The Logic of Scientific Discovery,* London: Hutchinson, 1958, repr. New York: Harper Torchbooks.

Popper, K. R. (1944–5) 'The poverty of historicism', *Economica* 11–12, revised and reprinted as *The Poverty of Historicism*, London: Routledge and Kegan Paul 1957, 2nd corr. ed. 1961.

Russell, B. (1945) 'Letter', *Journal of Education* 77: 224.

Uebel, T. E. (1994) 'The positivism dispute revisited', in *Institute Vienna Circle Yearbook* 2: 323–9.

Uebel, T. E. (1997) 'Sozialwissenschaft auf physikalistischer Grundlage im Wiener Kreis', in H.-J. Sandkühler (ed.) *Philosophie und Wissenschaften. Formen und Prozesse ihrer Interaktion*, Frankfurt a.M.: Peter Lang: 169–92.

Uebel, T. E. (1998) 'Enlightenment, positivism and reflexivity. the Vienna Circle's scientific world-conception', in A. O. Rorty (ed.) *Philosophy as Education*, London: Routledge, 1998: 418–38.

Urbach, P. (1978) 'Is any of Popper's arguments against historicism valid?', *British Journal for the Philosophy of Science* 29: 117–30.

Urbach, P. (1987) 'The scientific standing of evolutionary theories of society', *LSE Journal* 1: 23–42.

Watkins, J. W. N. (1953) '[Review of] Hayek, *The Counter-Revolution of Science*', *Ethics* 64, repr. in Wood and Woods 1991: 103–7.

Wood, J. C. and Woods, R. N. (eds.) (1991) *Friedrich A. Hayek: Critical Assessments vol. 2*, London: Routledge.

10 The rise of physicalism

David Papineau

> No one could seriously, rationally suppose that the existence of antibiotics or electric lights or rockets to the moon disproves . . . mind-body dualism.
>
> (Stephen R. L. Clark, 1996)

Introduction

In this paper I want to discuss the way in which physical science has come to claim a particular kind of hegemony over other subjects in the second half of this century. This claim to hegemony is generally known by the name of 'physicalism'. In this paper I shall try to understand why this doctrine has come to prominence in recent decades. By placing this doctrine in a historical context, we will be better able to appreciate its strengths and weaknesses.

As a preliminary, note that contemporary physicalism is an ontological rather than a methodological doctrine. It claims that everything is physically constituted, not that everything should be studied by the methods used in physical science.[1] This emphasis on ontology rather than methodology marks a striking contrast with the 'unity of science' doctrines prevalent among logical positivists in the first half of the century (and discussed by Thomas Uebel in the previous chapter of this book). The logical positivists were much exercised by the question of whether the different branches of science, from physics to psychology, should all use the same method of controlled observation and systematic generalization. They paid little or no attention to the question of whether everything is made of the same physical stuff.

By contrast, physicalism, as it is understood today, has no direct methodological implications. Some physicalists uphold the view that all sciences should use the 'positivist' methods of observation and generalization. But as many would deny this. You can be a physicalist about biology, say, and yet deny that biology is concerned with laws, or a physicalist about sociology, and yet insist that sociology should use the method of empathetic *verstehen* rather than third-person observation.

This methodological liberalism goes with the fact that the ontological claims of *fin-de-siècle* physicalism are often carefully nuanced. If physicalism simply meant type-type physical reduction, of the kind classically characterised in Ernst Nagel's *The Structure of Science* (1961), then methodological unity of science would arguably follow, in principle at least, from physicalism. But physicalism today clothes itself in various subtler shades. We have physical supervenience, physical realization, token-token physical identity, and so on. These more sophisticated doctrines leave plenty of room for different sciences to be studied in different ways.

But I am already drifting away from the main subject of this paper. My concern here is not to distinguish the different species of physicalism, though I shall touch on this in passing later, but to try to understand the reasons for physicalism of any kind. Why have so many analytic philosophers in the second half of the twentieth century suddenly become persuaded that everything is physical?

Fashions and arguments

It certainly was not always so. Perhaps the easiest way to highlight the recent shift in thinking about physicalism is to recall a once heated mid-century debate about the status of psychological explanation. In contemporary terms, this debate was about the scientificity of 'folk-psychology'. On the one side were those, like Carl Hempel and A. J. Ayer, who argued that 'reasons are causes'. By this they meant that psychological explanations are underpinned by empirical generalizations, implicit in everyday thought, which link psychological states like belief and desire to subsequent behaviour. Opposed to Hempel and Ayer were thinkers like William Dray and Peter Winch, who argued that the links between reason and action are 'logical' or 'meaningful', not empirical (Hempel 1942; Ayer 1969; Dray 1957; Winch 1958).

In one respect this old debate is still up-to-date. It concerned the question of whether everyday psychological thinking is suitable for incorporation in a scientific psychology – whether folk psychology is a 'proto-science', as it is sometimes put – and this question is still very much a live issue. But at another level the old debate is now quite outmoded. This is because the participants in the old debate showed little or no interest in the question of how the mind relates to the brain. They wanted to know whether there are testable, empirical laws linking mental states to behaviour. But they seemed to see no connection between this issue and the question of the relation of mental states to brain states. (In one perfectly good sense, they were addressing the issue of whether psychology is part of 'the proper ambition of science'. But for them this meant the question of whether categories like belief and desire conform to regular patterns at the psychological level, not the further question of how the categories of belief and desire relate to occurrences at the physical level.)

Nowadays, by contrast, everybody has a view on this latter question. Indeed nearly all analytic philosophers in this area, including those who side with Dray and Winch against the scientificity of common-sense psychology, now accept that the mind is in some way constitutively connected with the brain. (Thus consider Donald Davidson. He is the modern champion of the Dray-Winch view that the explanatory links between reason and action are a *sui generis* matter of rational understanding, not scientific law. Yet he made his name by arguing that, even so, 'reasons *are* causes'. In effect, his contribution was to show how the Dray-Winch methodological denial of psychological *laws* could be combined with a physicalist commitment to mind-brain *constitution* (Davidson 1963).

This transformation of the old 'reasons and causes' debate happened very quickly. Until the 1950s the issue was purely about lawlike patterns. The issue of mind-brain identity was not on the agenda. Then suddenly, in the 1950s and 1960s, a whole stream of philosophers came out in favour of physicalism. First there were Herbert Feigl and the Australian central state materialists, and they were followed in short order by Donald Davidson, David Lewis, and functional state theorists like Hilary Putnam. While the old 'reasons and causes' issue continued to be debated, from now on this debate took place within the larger context of physicalist assumptions about the mind-brain relation (Feigl 1958; Place 1956; Smart 1959; Armstrong 1968; Davidson 1963, 1970; Lewis 1966; Putnam 1960).

Why exactly did physicalism come to prominence in this way in the 1950s and 1960s? Those antipathetic to physicalism sometimes like to suggest that the emergence of physicalism is essentially a matter of fashion. On this view, the rise of physicalism testifies to nothing except the increasing prestige of physical science in the modern *Weltaunschang*. We have become dazzled by the gleaming status of the physical sciences, so the thought goes, and so foolishly try to make our philosophy in its image. (Thus Stephen Clark, in the sentence immediately following the quote at the begining of this paper: 'But such achievements [antibiotics, lights, rockets] lend authority to "science", and science . . . is linked in the public mind with atheistic materialism'(Clarke 1996).)

I think this attitude quite underestimates the significance of contemporary physicalism. What is more, it does not really answer the question about physicalism's sudden emergence. It is not as if the prestige of physics suddenly had a big boost in the middle of the twentieth century. I would say that physics has been pretty prestigious for about 300 years, with occasional ups and downs. Yet the philosophical physicalism we are concerned with is a distinctively late twentieth-century phenomenon.

In this paper I want to offer a different suggestion. My explanation for the rise of physicalism will be that it follows from an argument, or rather a family of arguments, the crucial premise of which was not available, at least to philosophers, until relatively recently. This is because this crucial premise is an empirical claim, and the evidence for it has only become clear-cut over the last century. Prior to that, this premise was not upheld by scientific theory, and so was unavailable as a basis for philosophical argument.

If this explanation is right, it casts a different light on physicalist views. Physicalism has been pressed on philosophers, not by fad or fashion, but by a newly available line of argument. In saying this, I do not want to suggest that the argument for physicalism is uncontroversial, or that the crucial premise I shall focus on is incontrovertible. But I do want to urge that physicalism deserves to be taken seriously, and that those who want to oppose it have an obligation to show where the argument in its favour goes wrong.

Of course, there are those, like Stephen Clark, who think that 'no one could seriously, rationally suppose' that empirical considerations could possibly yield a disproof of mind-body dualism (Clarke 1996). I shall not explicitly engage with this attitude in what follows, but shall merely invite those who find it plausible to consider the matter again at the end of this paper. Of course, to repeat a point just made, the empirically-based arguments in favour of physicalism are not incontestable. But, even so, it scarcely follows that you have to be unserious or irrational to suppose that they in fact succeed in establishing physicalism. Indeed it is my contention in this paper that a number of the most influential of late twentieth-century analytic philosophers have supposed just that.

Phenomenalism and physicalism

Before I give my own explanation for the rise of physicalism, in terms of the new availability of an empirical argument, let me quickly consider an alternative possible explanation, namely, that the rise of physicalism is simply the other side of the demise of phenomenalism.

No doubt there is something to this thought. Phenomenalism was the dominant metaphysical view among logical positivists and other scientifically-minded analytic philosophers in the first half of this century.

And there certainly isn't much room within phenomenalism to be a physicalist. If you think that everything, including physical stuff, is logically constituted out of mental items like sense data, then you would seem already to have ruled out the thought that mental items are in turn constituted by physical items.

Even so, I do not think this is a sufficient explanation for the rise of physicalism. For one thing, the rejection of phenomenalism doesn't yet explain the acceptance of physicalism. After all, you can deny phenomenalism without embracing physicalism. Indeed a significant number of contemporary philosophers do exactly that. These philosophers reject phenomenalism, but see no reason to privilege the physical among the different categories of things that exist, and so do not agree that everything is physically constituted.

Apart from this, there is the question of why phenomenalism died in the first place. This is of course a big subject, and any full answer would have to mention Wittgenstein's private language argument and Sellars' attack on givens. But I suspect that just as influential as these was the empirical argument for physicalism I am about to discuss. It is a simple argument, from uncomplicated empirical premises, and phenomenalists would have been as well-placed to appreciate its force as anybody else. If there is anything to this suggestion, then it wasn't so much that physicalism happened to fill the space created when phenomenalism left the stage. Rather the argument for physicalism was itself partially responsible for the overthrow of phenomenalism.

It is high time I described this empirically-based argument for physicalism. It is simple enough in outline. The crucial empirical premise is *the completeness of physics*, by which I mean that all physical effects are due to physical causes. And the argument is then simply that, if all physical effects are due to physical causes, then anything that has a physical effect must itself be physical.

The important point, for our purposes, is that the premise here, the completeness of physics, is a doctrine with a history. It was not always widely accepted. In particular, it was only after some decades of the present century that it became part of scientifically educated commonsense. This in turn was because evidence favouring this thesis did not start to emerge until the mid-nineteenth century, and did not become generally persuasive until much later. Once the thesis was widely accepted, however, its implications were obvious, and nearly all philosophers with some acquaintance with modern physical science became physicalists.

In the rest of this chapter I shall proceed as follows. First, in the next two sections, I shall get a bit clearer about what the completeness of physics says, and how different philosophers have used it to argue for physicalism.

In the following sections I shall then examine the history of this thesis, and in particular the reasons why it has come to be widely accepted nowadays, even though it was not always.

The completeness of physics and the argument for physicalism

Let me start by formulating a more precise version of the thesis of the completeness of physics:

> All physical effects are fully determined by law by a purely physical prior history.[2]

Note first that this thesis does not yet assert physicalism. Physicalism is the doctrine that everything, including *prima facie* non-physical stuff, is physical. But the completeness of physics does not itself say anything about non-physical things. It is purely a doctrine about the structure of the physical realm. It says that, if you start with some physical effect, then you will never have to leave the realm of the physical to find a fully sufficent cause for that effect.[3]

If we want to get from the completeness of physics itself to the imperialist phsyicalist conclusion that *everything* is physical, we need an argument. However, the general shape of such an argument is not hard to find. As I put it in the last section, if the completeness of physics is right, and all physical effects are due to physical causes, then anything *that has a physical effect* must itself be physical. Or, to put it the other way round, if the completeness of physics is right, then there is no room left for anything non-physical to make a difference to physical effects, so anything that does make such a difference must itself be physical.

Some version of this line of thought underlies the writings of all the philosophers who started arguing for physicalism in the 1950s and 1960s. Thus, for example, consider Smart's thought that we should identify mental states with brain states, for otherwise those mental states would be 'nomological danglers' which play no role in the explanation of behaviour. Similarly, reflect on Armstrong's and Lewis's argument that, since mental states are picked out by their causal roles, including their roles as causes of behaviour, and since we know that physical states play these roles, mental states must be identical with those physical states. Or, again, consider Davidson's argument that, since the only laws governing behaviour are those connecting behaviour with physical antecedents, mental events can only be causes of behaviour if they are identical with those physical antecedents.

There is much to say about these arguments, and I shall say some of it later in the chapter. But the point I want to make here is that none of these arguments would seem even slightly plausible without the assumption of the completeness of physics. To see this, imagine that the completeness of physics were not true, and that some physical effects (the movements of arms, perhaps, or the firings of the motor neurones which instigate those movements) were not determined by law by prior physical causes at all, but by *sui generis* non-physical mental causes, such as decisions, say, or exercises of will, or perhaps just pains. Then, first, *contra* Smart (1959), mental states would not be 'nomological danglers', but directly efficacious in the production of behaviour; second, *contra* Armstrong (1968) and Lewis (1980), it would not necessarily be physical states which played the causal roles by which we pick out mental states, but quite possibly the *sui generis* mental states themselves; and third, *contra* Davidson (1970), it would not be true that the only laws governing behaviour are those connecting behaviour with physical antecedents, since there would also be laws connecting behaviour with mental antecedents.[4]

Comments on the causal argument for physicalism

The interesting historical question, to which I shall turn shortly, is why these completeness-of-physics-based arguments started appearing when they did. But first it will be useful to clear away a bit of philosophical undergrowth. Those readers who are more interested in history than philosophical niceties may wish to skip ahead to the next section.

There are significant differences between the completeness-based arguments put forward by Smart, Armstrong, Lewis, and Davidson and other physicalist writers. However, rather than getting entangled in detailed comparisons, let us focus on one canonical form of this argument, which I shall call the 'causal argument' (Crane 1995, Sturgeon 1998). This will enable me to make some general structural points.

Premise 1 (the completeness of physics):
All physical effects are fully determined by law by a purely physical prior history.

Premise 2 (causal influence):
All mental occurrences have physical effects.[5]

Premise 3 (no universal overdetermination):
The physical effects of mental causes are not all overdetermined.

Conclusion:
Mental occurrences must be identical with physical occurrences.

Let me add some comments.

First, regarding *the ontology of causes*. The force of this causal argument is extremely sensitive to how you think about causation. If, like Donald Davidson (1980), you think of the relata of causation as events, and think of events in turn as basic particulars, then the argument concludes only that mental and physical descriptions pick out the same *events*, not that there is any constitutive relationship between mental and physical *properties*. On the other hand, if you think of the relata of causation as instantiations of properties, or more generally as facts (Mellor 1995), then the argument promises to establish the stronger conclusion that mental properties are identical with physical ones. Since the stronger version is the more interesting, and since facts in any case seem to me the better candidates for the relata of causation, I shall read the argument in this way henceforth.

Second, regarding *accepting overdetermination*. The causal argument seems pretty clearly to be valid.[6] So those who reject the conclusion must reject one of the three premises. All three moves are found in the literature. The status of premise one, the completeness of physics, will occupy most of what follows. This leaves premises two and three. Let us first consider rejecting premise three, the premise of no universal overdetermination.

To reject this presmise is to accept that the physical effects of mental causes are always overdetermined. This 'belt and braces' view is defended by Gabriel Segal and Elliott Sober (1991) and D. H. Mellor (1995, pp. 103–5). In response to the worry that this view seems to imply that your arm would still have moved even if you had not felt a pain (because your C-fibres would still have fired, say), these philosophers argue that the distinct mental and physical causes may themselves be strongly counterfactually dependent. Still, this then raises the question of *why* such causes should always be so counterfactually dependent, if they are genuinely distinct. Possible causal mechanisms underpinning this dependence can be imagined, but there seems to me no good reason to believe in them.

Third, regarding *epiphenomenalism and pre-established harmony*. What about premise two? The possibility of denying this premise is familiar enough, under the guise of 'epiphenomenalism' or 'pre-established harmony'. If you are prepared to accept that mental states do not have physical effects, and are indeed 'nomological danglers' with respect to the causation of behaviour, then the above argument for physicalism will not move you, for you will not embrace its second premise. I leave it to readers to decide whether this denial of the efficacy of the mental is a price worth paying to avoid physicalism.[7]

While we are on this point, it is worth noting that one of the most popular versions of physicalism, namely functionalism, is arguably a closet version of epiphenomenalism. By functionalism I mean the view that identifies a mental state with a 'second-order state', that is, the state-of-having-some-state-which-plays-a-certain-role, rather than with the first-order physical state which actually plays that role. Since the second-order mental state cannot be *identified* with the first-order physical state (rather, it is 'realised' by it), it is not clear that it can be deemed to cause what that first-order state causes, such as items of behaviour. So functionalism threatens the epiphenomenalist denial of premise two, the claim that mental states have physical effects.

The recognition of this difficulty has put functionalism under some pressure recently. One option is to turn away from functionalism, and insist that mental states are first-order states after all, and so strictly identical with physical states (Lewis 1980). This option in effect upholds a strong version of premise two, and allows it to argue for the full identity of mental with physical properties.

Another option is to read 'causation' generously in premise two, and assume only that mental states cause physical effects in the weaker sense that either they cause them directly or they have realizers that cause them directly. That is, we might allow a state to 'cause' in virtue of having a realizer which causes. If we do this, then the causal argument will no longer require us to identify conscious states with strictly physical states, but it will still give us an argument for identifying them with second-order states which are physically realized. For unless we suppose this identification, even the weaker version of premise two will force us to the unhappy conclusion that behavioural effects are overdetermined by two ontologically quite unrelated causes.[8]

Fourth, regarding *non-causal realms*. This discussion of epiphenomenalism shows that the causal argument for physicalism only applies to non-physical occurrences that *do* have physical effects. Without premise two, there is no argument, since it is only on the assumption that the non-physical occurrences in question are *not* 'causal danglers' that we need to identify them with something physical.

This shows that there are limits to this form of argument for physicalism. At the beginning of this paper I characterised physicalism as the doctine that 'everything is physically consituted'. However, this ambitious claim outstrips anything that can be delivered by the causal argument. For the causal argument has no grip on putative realms of reality that are outside the causal realm altogether, and so *a fortiori* do not have physical effects. I particularly have in mind here the realms of mathematics, and of moral and other values. While some philosophers have supposed that mathematical or moral facts do have physical effects, this is not the normal way to think about them. And, if we do deny that moral or mathematical facts have physical effects, then our causal argument will provide no basis for identifying them with physical facts.[9]

I myself think that this limitation to the causal argument constitutes a genuine boundary to the proper ambitions of physicalism. I think that physicalism is best formulated, not as the claim that everything is physical, but as the significantly weaker claim that everything which interacts causally with the physical world is physical. This leaves it open that there may be non-causal realms of reality which are not physically constituted, such as the realm of moral worth, or of beauty, or of mathematical objects.

Of course, there may be other problems with such non-physical realms. For example, it is not clear how we may come by knowledge of such realms, if they can have no physical effects on our sense organs. But these further arguments are by no means clear-cut, and there is no special reason why they should be accepted by everybody who accepts the causal argument. Because of this, I shall use 'physicalism' in the rest of this paper specifically for the doctrine that everything with physical influence is physical, whatever may be true of other realms.

Fifth, regarding *what physics is.* In a moment I shall turn to the history of the completeness of physics. But first we need to address a terminological issue, one that may have been worrying readers for some time. How exactly is 'physics' to be understood in this context of the causal argument? An awkward dilemma may seem to face anyone trying to defend the crucial first premise, the completeness of physics. If we take 'physics' to mean the subject matter currently studied in departments of physics, discussed in physics journals, and so on, then it seems pretty obvious that physics is not complete. The track record of attempts to list *all* the fundamental forces and particles responsible for physical effects is not good, and it seems highly likely that future physics will identify new categories of physical cause. On the other hand, if we mean by 'physics' the subject matter of such future scientific theories, then we seem to be in no position to assess its completeness, since we do not yet know what it is.

This difficulty is more apparent than real. If you want to use the causal argument, it is not crucial that you know exactly what a complete physics would include. Much more important is to know what it will not include. Suppose, for example, that you have an initial idea of what you mean by 'mental' (the sentient, say, or the intentional, or perhaps just whatever events occur specifically in the heads of intelligent beings). And suppose now that you understand 'physical' as simply meaning 'non-mental', that is, as standing for those properties which can be identified without using this specifically mental terminology. Then, provided we can be confident that the 'physical' in this sense is complete, that is, that every non-mental effect is fully determined by *non-mental* antecedents, then we can conclude that all mental states must be identical with something non-mental (otherwise mental states could not have non-mental effects). This understanding of 'physical' as 'non-mental' might seem a lot weaker than most pre-theoretical understandings, but note that it is just what we need for philosophical purposes, since it still generates the worthwhile conclusion that the mental must be identical with the non-mental; given, that is, that we are entitled to assume that the non-mental is complete.

The same point applies if we want to apply the causal argument to chemical, or biological, or economic states. As long as we can be confident that all non-chemical effects are fully caused by non-chemical (non-biological/non-economic . . .) states, then we can conclude that all chemical (biological/economic . . .) states must be identical with something non-chemical (non-biological/non-economic . . .).

We might not know enough about physics to know exactly what physics does include. But as long as we are confident that it *excludes* such-and-such special categories, then we can use the causal argument to conclude that these special categories are in fact identical with other kinds. I shall suppose this indirect understanding of 'physics' in what follows: it should simply be understood as that set of properties which can be specified without appeal to whichever special vocabularies (mental, biological . . .) we are interested in. Correspondingly, the completeness of physics will be the doctrine that such non-special effects are always fully accounted for by non-special causes (cf. Papineau and Spurrett, 1999).

Descartes and Leibniz

Let us now concentrate on the history of the completeness of physics. The important question, as we have just seen, is whether any non-special effects are produced by *sui generis* special causes. True, the exact content of this question will be relative to which special categories we are interested in, for the reasons just explained. Still, we can take it for the moment that we are interested in a relatively strong version of the completeness of physics, and in particular one which would rule out *sui generis* mental causes, along with biological, economic, social, or other even more special causes. So let us focus for now on the question of whether there are any non-mental effects which cannot be accounted for without reference to *sui generis* mental causes.

When I first became interested in the causal argument a few years ago, I recognised that there were many points where it could be queried. However, I assumed that the completeness premise itself was quite uncontentious. Surely, I thought, everybody agrees that the movements of matter, such as the movements of molecules in your arm, can in principle always be fully accounted for in terms of prior physical causes, such as physical activity in your nerves, which in turn is due to physical activity in your brain, and so on.

To my surprise, I discovered that some people did not agree. They did not see why some physical occurrences, in our brains perhaps, should not have irreducibly mental causes. My first response, when presented with this thought, was to attribute it to an insufficient education in the physical sciences. Sometimes I went so far as to communicate this diagnosis to those who disagreed with me. However, when they then asked me, not unreasonably, to show them where the completeness of physics is written down in the physics textbooks, I found myself in some embarrassment. Once I was forced to defend it, I realised that the completeness of physics is by no means self-evident. Indeed further reading has led me to realize, far from being self-evident, it is an issue on which the post-Galilean scientific tradition has changed its mind several times.

My original thought was that the completeness of physics would follow from the fact that physics can be formulated in terms of conservation laws. If the laws of mechanics tell us that important physical quantities are conserved whatever happens, then does it not follow that the later states of physical systems are always fully determined by their earlier physical states?

Not necessarily. It depends on what conservation laws you are committed to. Consider Descartes' mechanics. This incorporated the conservation of what Descartes called 'quantity of motion', by which he meant mass times speed. That is, Descartes held that the total mass times speed of any isolated collection of bodies is guaranteed to remain constant, whatever happens to them. However, this alone does not guarantee that physics is complete. In particular, it does not rule out the possibility of physical effects that are due to irreducibly mental causes.

This is because Descartes' *quantity of motion* is a non-directional (scalar) quantity, defined in terms of speed, as opposed to the directional (vectorial) Newtonian notion of linear *momentum*, defined in terms of *velocity*. Because of this, the *direction* of a body's motion can be altered without altering its quantity of motion. As Roger Woolhouse explains the point, in an excellent discussion of the relevance of seventeenth-century mechanics to the mind-brain issue (Woolhouse 1985), a car rounding a corner at constant speed conserves its 'quantity of motion', but not its momentum.

This creates room for *sui generis* mental causes to alter the *direction* of a body's motion without violating Descartes' conservation principle. Descartes' conservation principle means does mean that, if one physical body starts going faster, this must be due to another physical body going slower. But his principle does not require that, if a physical body changes direction, this need result from any other physical body changing direction.

Even if the change of direction results from an irreducibly mental cause, the quantity of motion of the moving body remains constant.

According to Leibniz, Descartes exploited this loophole to explain how the mind could affect the brain. As Leibniz tells the story, Descartes believed that the mind nudges moving particles of matter in the pineal gland, causing them to swerve without losing speed, like the car going round the corner. This then explained how the mind could affact the brain without violating the conservation of 'quantity of motion' (Leibniz 1898 [1696]: 327).

Now, there is little evidence that Descartes actually saw things this way, nor indeed that he was particularly worried about how the laws of physics can be squared with mind-brain interaction. Still, whatever the truth of Leibniz's account of Cartesian theory, his next point deserves our attention. For Leibniz proceeds from his analysis of Descartes to the first-order assertion that the *correct* conservation laws, unlike Descartes' conservation of quantity of motion, *cannot* in fact be squared with mind-body interaction.

Leibniz's conservation laws were in fact a great improvement on Descartes'. In place of Descartes' conservation of 'quantity of motion', Leibniz upheld both the conservation of linear *momentum* and the conservation of kinetic *energy*. These two laws led him to the correct analysis of impacts between moving bodies, a topic on which Descartes had gone badly astray. And, in connection with our present topic, they persuaded him that there is no room whatsoever for mental activity to influence motion of matter.

In effect, the conservation of linear momentum and of kinetic energy together squeeze the mind out of the class of events that cause changes in motion. Leibniz's two conservation laws, plus the standard seventeenth-century assumption of no physical action at a distance, are themselves sufficient to fix the evolution of all physical processes. The conservation of momentum requires the preservation of the same total amount of quantity of motion in *any given direction*, thus precluding any possibility of mental nudges altering the direction of moving physical particles. Moreover, the conservation of energy, when added to the conservation of momentum, fully fixes the speed and direction of impacting physical particles after the collide.[10] So there is no room for anything else, and in particular for anything mental, to make any difference to the motions of physical particles, if Leibniz's two conservation laws are to be respected.

We can simplify the essential point at issue here by noting that Leibniz's conservation laws, unlike Descartes', ensure physical determininism. They imply that the physical states of any system of bodies at one time fix their state at any later time. Physical determinism in this sense is certainly sufficient for the completeness of physics, even if the possibility of quantum-mechanical indeterminism means it is not necessary (cf. note 2).

So Leibniz's dynamics, unlike Descartes', makes it impossible for anything except the physical to make a difference to anything physical.

Leibniz was fully aware of the implications of his dynamical theories for mind-body interaction (cf. Woolhouse 1985). However he did not infer mind-brain identity from his commitment to the completeness of physics. Instead he adopted the doctine of pre-established harmony, according to which the mental and physical realms are each causally closed, but pre-arranged by the divine will to march in step in such a way as to display the standard mind-brain correlations. In terms of the canonical causal argument laid out earlier, Leibniz is here denying premise two, about the causal influence of mind on matter. He avoids identifying mental causes with physical causes, in the face of the completeness of physics, by denying that mental causes ever have physical effects.

Newtonian physics

Some readers might wonder why this is not the end of the issue. Given that Leibniz established, against Descartes, that both momentum and energy are conserved in systems of moving particles, then why was the history of the mind-brain argument not already over? Of course, we might not nowadays want to follow Leibniz in opting for pre-established harmony, as opposed to simply embracing mind-brain identity. But this is simply because we favour a different response to the causal argument laid out earlier, not because we have any substantial premises Leibniz lacked. In particular, the crucial first premise of the causal argument, the completeness of physics, would seem already to have been available to Leibniz. So does this not mean that everything needed to appreciate the causal argument was already to hand in the second half of the seventeenth century, long before the rise of twentieth-century physicalism?

It was, but only on the assumption Leibniz gives us the correct dynamics. However, Leibniz's physical theories were quickly eclipsed by those of Newton, and this then re-opened the whole issue of the completeness of physics.

The central point here is that Newton allowed forces other than impact. Leibniz, along with Descartes and all other pre-Newtonian proponents of the 'mechanical philosophy', took it as given that all physical action is by contact. They assumed that the only possible cause of a change in a physical body's motion is the impact of another physical body. (Or more precisely, as we are telling the story, Descartes supposed that the only possible *non-mental* cause of physical change is impact, and Leibniz then argued that *mental* causes other than impact are not possible either, if the conservation of momentum and energy are to be respected.)

Newtonian mechanics changed the whole picture. This is because Newton did not take impact as his basic model of dynamic action. Rather his basic notion is that of an *impressed force*. Rather than thinking of 'force' as something inside a body which might be tranferred to other bodies in impact, as did all his contemporaries (and indeed as did most of his successors for at least a century), Newton thought of forces as disembodied entities, acting on the affected body from outside.[11] An impressed force 'consists in the action only, and remains no longer in the body when the action is over.' Moreover, 'impressed forces are of different origins, as from percussion, from pressure, from centripetal force.' (Newton 1960 [1686], Definition IV.) Gravity was the paradigm. True, the force of gravity always arose from the presence of massive bodies, but it pervaded space, waiting to act on anything that might be there, so to speak, with a strength as specified by the inverse square law.

Once disembodied gravity was allowed as a force distinct from the action of impact, then there was no principled barrier to other similarly disembodied special forces, such as chemical forces, or magnetic forces, or forces of cohesion (cf. Newton, 1952 [1704], Queries 29–31) – or indeed vital and mental forces.

Nothing in classical Newton thinking rules out special mental forces. While Newton has a general law about the effects of his forces (that they cause proportional changes in the velocities of the bodies they act on), there is no corresponding general principle about the causes of such forces. True, gravity in particular is governed by the inverse square law, which fixes gravitional forces as a function of the location of bodies with mass. But there is no overarching principle dictating how forces in general arise. This opens up the possibility that there may be *sui generis* mental forces, which would mean that Newtonian physics, unlike Leibnizian physics, is not physically complete. Some physical processes could have non-physical mental forces among their causal antecedents.[12]

The switch from a pure impact-based mechanical philosophy to the more liberal world of Newtonian forces thus undermined Leibniz's argument for the completeness of physics. Leibniz could hold that the principles governing the physical world leave no room for mental acts to make a difference because he had a simple mechanical picture of the physical world. Bodies preserve their motion in any given direction until they collide, and then they obey the laws of perfect elastic impact. The Newtonian picture is far less pristine, and gives no immediate reason to view physics as complete.

You might that think the conservation laws of Newtonian physics would themselves place constraints on the generation of forces, in such a way as to restore the completeness of physics. But this would be a somewhat anachronistic thought. Conservation laws did not play a central role in Newtonian thinking, at least not in that of Newton himself and his immediate followers. True, Newton's mechanics does imply the conservation of *momentum*. This falls straight out of his Third Law which requires that 'action and reaction' are always equal. But it is a striking feature of Newtonian dynamics that there is no corresponding law for energy.[13]

Of course, as we shall see in the next section, the principle of the conservation of kinetic *and* potential energy in all physical processes did *eventually* become part of the Newtonian tradition, and this does impose a general restriction on possible forces, a restriction expressed by the requirement that all forces should be 'conservative'. But this came much later, in the middle of the nineteenth century, and so had no influence on the range of possible forces admitted by seventeenth or eighteenth-century Newtonians. (Moreover, it is a nice question, to which we shall return at length below, how far the principle of the conservation of kinetic plus potential energy, with its attendant requirement that all forces be conservative, does indeed constitute evidence against *sui generis* mental forces.)

In any case, whatever the significance of later Newtonian derivations of the conservation of energy, early Newtonians themselves certainly saw no barrier to the postulation of *sui generis* mental forces. In a moment I shall give some examples. But first it will be helpful to distinguish in the abstract two ways in which such a Newtonian violation of the completeness of physics could occur.

First, and most obviously, it could follow from the postulation of *indeterministic* mental forces. If the determinations of the self (or of the 'soul', as they would have said in the seventeenth and eighteenth centuries) could influence the movements of matter in spontaneous ways, then the world of physical causes and effects would obviously not be causally closed, since these spontaneous mental causes would make a difference to the unfolding of certain physical processes.

Second, it is not even necessary for the violation of completeness that such *sui generis* special forces operate indeterministically. Suppose that the operation of mental forces were governed by fully *deterministic* force laws (suppose, for example, that mental forces obeyed some inverse square law involving the presence of certain particles in the brain). Then mental forces would be part of Newtonian dynamics in just the same sense as gravitational or electrical forces: we could imagine a system of particles evolving deterministically under the influence of all these forces, including mental forces, with the forces exerted at any place and time being deterministically fixed by the relevant force laws. Even so, this deterministic model would still constitute a violation of the completeness of physics, for the physical positions of the particles would depend *inter alia* on prior mental causes, and not exclusively on prior physical causes.

Did I not say at the end of the last section that determinism is sufficient for the completeness of physics (even if not necessary, because of quantum mechanics)? No. What I said was that *physical* determinism (the doctrine that prior *physical* conditions alone are enough to determine later physical conditions) is sufficient for the completeness of physics. However, we can accept determinism as such without accepting physical determinism, and so without accepting the completeness of physics. In particular, we can have a deterministic model in which mental forces play an essential role, and in which the physical sub-part is therefore not causally closed.

You might feel (indeed might have been feeling for some time) that a realm of deterministic mental forces would scarcely be worth distinguishing from the general run of physical forces, given that they would lack the spontaneity and creativity that is normally held to distinguish the mental from the physical. And you might think that it is therefore somewhat odd to view them as violating the completeness of physics. I happily concede that there is something to this thought. But I would still like to stick to my terminology, as stipulated at the end of section 5, which assumed an initial sense for 'mental' (as sentient, or intentional, or intelligent), and then defined the 'physical' as whatever can be identifed without alluding to such mental properties, which then makes even deterministically governed sui generis mental forces come out 'non-physical', since they cannot be so non-mentally identified. This is the terminology which best fits with our original interest in the causal argument for physicalism. We do not want deterministic mental forces to be counted as consistent with the 'completeness of physics', precisely because *this* kind of 'completeness of physics' would not be any good for the causal argument: if mental forces are *part* of what makes 'physics' complete, then we will not be able to argue from this that mental forces must be identical with some *other* (non-mental) causes of their effects.

So far I have merely presented the possibility of special Newtonian forces as an abstract possibility. However, the postulation of such forces was a commonplace among eighteenth-century thinkers, particularly among those working in anatomy and physiology. Many of the theoretical debates in these areas were concerned with the existence of vital and mental forces, and with the relation between them. Among those who debated these issues, we can find both the indeterministic and deterministic models of mental forces.[14]

Thus consider the debate among eighteenth-century physiologists about the relative roles of the forces of *sensibility* and *irritability*. This terminology was introduced by the leading German physiologist Albrecht von Haller, Professor of Anatomy at Göttingen from 1736. Haller thought of 'sensibility' as a distinctively mental force. 'Irritability' was a non-mental but still peculiarly biological power. ('What should hinder us from granting irritability to be a property of the animal *gluten*, the same as we acknowledge gravity and attraction to be properties of matter in general?' (Haller, 1936 [1751]).) Haller took the force of sensibility to be under the control of the soul and to operate solely through the nerves. Irritability, by contrast, he took to be located solely in the muscle fibres.

In distinguishing the mentally directed force of sensibility from the more automatic force of irritability, Haller can here be seen as conforming to my model of *indeterministic* mental forces. Where the force of irritability is determined by prior stimuli and is independent of mental agency, the force of sensibility responds to the spontaneous commands of the soul.

Haller's model was opposed by Robert Whytt (1714–66) in Edinburgh. In effect Whytt can be seen as merging Haller's distinct vital and mental forces, irritability and sensibility. On the one hand, Whytt gave greater power to the soul: he took it that a soul or 'sentient principle' is distributed throughout the body, not just in the nerves, and is responsible for all bodily activities, from the flow of blood and motion of muscles, to imagination and reasoning in the brain. But at the same time as giving greater power to this sentient principle, he also rendered its operations *deterministic*. He explictly likened the sentient principle to the Newtonian force of gravity, and viewed it as a necessary principle which acts according to strict laws. Whytt can thus be seen as exemplifying my model of deterministic mental forces: the sentient principle is simply another deterministic Newtonian force, just like gravity and the others, in that its operations are fixed by a definite force law (Whytt 1755).

The conservation of energy

In this section I want to consider how the principle of the conservation of energy eventually emerged within the tradition of Newtonian mechanics, and how this bears on the completeness of physics. It will be useful to separate some different aspects of this emergence.

Rational mechanics

Through the eighteenth and early nineteenth centuries a number of mathematician-physicists, among the most important of whom were Jean d'Alembert (1717–83), Joseph Louis Lagrange (1736–1813), the Marquis de Laplace (1749–1827) and William Hamilton (1805–65), developed a series of mathematical frameworks designed to simplify the analysis of the motion of interacting particles. These frameworks allowed physicists to abstract away from detailed forces of constraint, such as the forces holding rigid bodies together, or the forces constraining particles to move on surfaces, and concentrate on the effects produced by other forces. (See Elkana 1974 chapter II for the history, and Goldstein 1964 for the mathematics.)

These mathematical developments also implied that, under certain conditions, the sum of kinetic energy and potential energy remains constant. Roughly, when all forces involved are independent of the velocities of the interacting particles and of the time (let us call forces of these kinds 'conservative'), then the sum of actual kinetic energy (measured by $\frac{1}{2} \Sigma mv^2$) plus the potential to generate more such energy (often called the 'tensions' of the system) is conserved: when the particles slow down, this builds up 'tensions', and, if those 'tensions' are expended, the particles will speed up again.

We now think of this as the most basic of all natural laws. But this attitude was no part of the original tradition in rational mechanics. There were two reasons for this. First, the Newtonian scientists in this tradition were not looking for conserved quantities anyway. As I explained earlier, conservation principles played little role in classical Newtonian thinking. True, Leibniz himself had urged the conservation of kinetic energy (under the guise of '*vis viva*'), but by the eighteenth century Leibniz's influence had been largely eclipsed by Newton's. Second, the conservation of potential and kinetic energy in any case only holds under the assumption that all forces are conservative. We nowadays take this requirement to be satisfied for all fundamental forces. But this again was no part of eighteeenth-century thinking. Some familiar forces happen to be conservative, but plenty of other forces are not. Gravitation, say, is conservative, since it depends only on the positions of the particles, and not on their velocities, nor the elapsed time. But, by contrast, frictional forces are not conservative, since they depend on the velocity of the decelerated body relative to the medium. And correspondingly frictional forces do not in any sense seem to conserve energy: when they decelerate a body, no 'tension' is apparently built up waiting to accelerate the body again.

For both these reasons, the tradition in rational mechanics did not initially view the conservation of kinetic and potential energy in certain systems as of any great significance. On the contrary, it was simply a handy mathematical consequence which falls out of the equations when the operative forces all happen to fall within a subset of possible forces (cf. Elkana 1974 chapter 2).

Equivalence of heat and mechanical energy

In the first half of the nineteenth century a number of scientists, most prominently James Joule (1819–89), established the equivalence of heat and mechanical energy, in the sense of showing that a specific amount of heat will always be produced by the expenditure a given amount of mechanical energy (as when a gas is compressed, say), and vice versa (as when a hot gas drives a piston).

These experiments suggested directly that some single quantity is preserved through a number of different natural interactions. They also had a less direct bearing on the eventual formulation of the conservation of energy. They indicated that apparently non-conservative forces like friction and other dissipative forces need not be non-conservative after all, since the kinetic energy apparently lost when they act will in fact be preserved by the heat energy gained by the resisting medium.[15]

The stage was now set for the formulation of a universal principle of the conservation of energy. We can distinguish three elements which together contributed to the formulation of this principle. First, the tradition of rational mechanics provided the mathematical scaffolding. Second, the experiments of Joule and others suggested that different natural processes all involve a single underlying quantity which could manifest itself in different forms. Third, these experiments also suggested that apparently non-conservative forces like friction were merely macroscopic manifestations of more fundamental conservative forces.

Of course, it is only with the wisdom of hindsight that we can see these different strands as waiting to be pulled together. At the time they were hidden in abstract realms of disparate branches of science. It took the genius of the young Hermann von Helmholtz (1821–94) to see the connections. In 1847, at the age of twenty-six, he published his monograph *Uber die Erhaltung die Kraft* ('On the conservation of force'). The first three sections of this treatise are devoted to the tradition of rational mechanics, and in particular to explaining how the total mechanical energy (kinetic plus potential energy) in a system of interacting particles is constant in those cases where all forces are familiar 'central forces' independent of time and velocity. The fourth section describes the equivalence between mechanical 'force' and heat, referring to Joule's results, while the last two sections extend the discussion to electric and magnetic 'forces', showing again that there are fixed equivalences between these 'forces', heat, and mechanical energy.[16]

Physiology

At the end of his treatise Helmholtz touches on the conservation of energy in living systems. Helmholtz was in fact a medical doctor by training, and had been a student in the Berlin physiological laboratory of Johannes Müller in the early 1840s, along with Emil Du Bois-Reymond (1818–96) and Ernst Brücke (1819–92). Together these students were committed to a reductionist programme in physiology, aiming to show that phenomena like respiration, animal heat, and locomotion could all be understood to be governed by the same laws as operate in the inorganic realm.

This physiological context undoubtedly played a fundamental role in Helmholtz's articulation of a universal principle of the conservation of energy. Because of his physiological interests, Helmholtz was interested in a principle that would cover *all* natural phenomena, including those in living systems, and not just such manifestly physical phenomena as mechanical motion, heat, and electromagnetism. Thus he took the crucial step of asserting that *all* forces conserve the sum of kinetic and potential energy; superficially non-conservative forces like friction are simply macroscopic manifestations of more fundamental forces which preserve energy at the micro-level. This then enabled Helmholtz to view the equivalences established by experimentalists like Joule, not just as striking local regularities, but as necessary consequences of a fundamental principle of mechanics. All natural processes must respect the conservation of energy, including processes in living systems.

It is noteworthy that neither the experimentalists like Joule, nor the mathematician-physicists in the rational mechanics tradition, made this crucial step to a universal principle. None of the scientists working experimentally on numerical equivalences between different processes, like Joule, generalised their discoveries into the claim that there is one quantity, energy, preserved in all natural interactions whatsoever. While it is true that a number of different scientists at the time were investigating such numerical equivalences (thus the historical thesis of the 'simultaneous discovery' of the conservation of energy), there is no reason to suppose that these scientists were generally inspired by any vision of the underlying unity of different natural processes. Similarly, there was nothing to attract mathematical physicists in the tradition of rational mechanics to the conclusion that all forces are conservative, for the reasons given above. They simply thought of such forces as the mathematically tractable special case where changes in kinetic and potential energy happen always to balance out.

Without the desire to bring living systems under a unified science, none of these scientists had any motive for synthesizing the different strands pulled together by Helmholtz. It was Helmholtz's combination of physiological interests and sophisticated physical understanding that precipitated the crucial step. He saw that, if we assume that all fundamental forces are conservative, then this guarantees that a certain quantity, the total energy, will be preserved in all natural processes whatsoever, including the organic processes that formed the focus of his interest.

Vital forces

Helmholtz was part of a tradition in experimental physiology which set itself in opposition to the previous generation of German *Natuurphilosophen*. During the eighteenth century the Newtonian categories of 'irritability' and 'sensibility' had gone through various transformations, and by the end of the century were widely referred to under the heading of 'Lebenskraft' or 'vital force', though there was continued disagreement on the precise nature of such forces. Meanwhile, within the tradition of German idealism, the notion of vital force had broken loose from its original Newtonian moorings, and became part of a florid metaphysics imbued with romanticism and idealism.

According to the *Natuurphilosophen*, organic matter was infused with a special power which organized and directed it. Following Blumenbach and Kant, Schelling took up the term '*Bildungstrieb*' ('formative drive'), because of what he saw as the excessively mechanical connotations of '*Lebenskraft*'. Schelling and the other *Natuurphilosophen* viewed this formative drive as having a quasi-mental aspect, which enabled it to to mediate between the 'archetypical ideas' or 'essences' of different species and the development of individual organisms towards that ideal form (see Coleman 1971 chapter 3, Steigerwald 1998).

The experimental tradition which included Helmholtz can be seen as a reaction to these extreme doctrines. However, it is striking that many of those associated with this tradition, though not Helmholtz himself, continued to admit the possible existence of vital forces, both before and after the emergence of the conservation of energy. This is less puzzling than it may at first seem. These physiological thinkers did not think of vital forces as the mystical intermediaries of the *Natuurphilosophen*, imbued with all the powers of creative mentality. Rather these thinkers were reverting to the tradition of eighteenth-century physiology. They simply viewed vital forces as special Newtonian forces, additional to gravitational forces, chemical forces and so on, and which happen to arise specifically in organic contexts. Justus von Leibig (1803–73), the leading physiological chemist of the time, and Müller, Helmholtz's own mentor, are clear examples of experimental physiologists who were prepared to countenance vital forces in this sense (cf. Coleman 1971 chapter VI, Elkana 1974 chapter IV).

Does the conservation of energy rule out vital (and mental) forces?

The interesting question, from the point of view of this chapter, is how far this continuing commitment to vital forces is consistent with the doctrine of the conservation of energy. There is certainly some tension between the two doctrines. It is noteworthy that Helmholtz himself, and his young colleagues from Müller's laboratory, were committed to the view that no forces operated inside living bodies that were not also found in simpler physical and chemical contexts (Coleman 1971: 150–4). Even so, there is no outright inconsistency between the conservation of energy and vital forces, and many late nineteenth-century figures were quite explicit, not to say enthusiastic, about accepting both.

In order to get clearer about the room left for vital (or mental) forces by the conservation of energy, recall how I earlier distinguished two ways in which early Newtonian theory left room for such special forces to violate the completeness of physics. First, such forces might operate spontaneously and indeterministically: nothing in early Newtonian theory would seem to rule out spontaneous forces ungoverned by any deterministic force law. Second, even if the relevant forces are governed by a deterministic force law, they may still be *sui generis*, in the sense that they may be distinct from gravitional forces, chemical forces, and so on, and may arise specifically in living systems or their brains.

The conservation of energy bears differentially on these two kinds of special forces. It does seem inconsistent with the first, spontaneous, kind of special force, but it does not directly rule out the second, deterministic kind.

Why should the conservation of energy rule out even a spontaneous special force? (Think of a spontaneous mental force that accelerates molecules in the pineal gland, say.) Why should such a force not simply respect the conservation of energy, by not causing accelerations which will violate it? But this does not really make sense. The content of the principle of the conservation of energy is that losses of kinetic energy are compensated by build-ups of potential energy, and vice versa. But we could not really speak of a 'build-up' or 'loss' in the potential energy associated with a force, if there were no force law governing the deployment of that force. So the very idea of potential energy commits us to a law which governs how the relevant force will cause accelerations in the future.

However, nothing in this argument rules out the possibility of vital, mental, or other special forces which *are* governed by deterministic force laws. After all, the conservation of energy in itself does not tell which basic forces operate in the physical universe. Are gravity and impact the only basic forces? What about electromagnetism? Nuclear forces? And so on. Clearly the conservation of energy as such leaves it open exactly which basic forces exist. It only requires that, whatever they are, they operate conservatively.

The death of emergentism

So a commitment to the conservation of energy by no means settled the question of whether *sui generis* mental or vital forces should be rejected and physics declared complete. True, some few thinkers, like Helmholtz himself, conjoined the conservation of energy with a denial of such special forces. But this was by no means mandated by the conservation of energy itself, for the reasons I have just explained. Accordingly, many other thinkers in the late nineteenth and early twentieth centuries took the opportunity to posit special forces of the kind allowed by the conservation of energy. So I still owe an explanation of what finally created a scientifically informed consensus against such special forces.

The issue is not straightforward, and there is no question of dealing with it fully here. But in this final section I would like to offer some outline conjectures. I shall proceed as follows. First, I shall take it as given that the conservation of energy at least was a settled doctrine. Of course there is a story to be told about this as well. But, for whatever reasons, the doctrine of the conservation of energy did win widespread acceptance within a decade or two of its initial formulation, and certainly none of the developments I am about to consider questioned its validity. Second, I shall lump mental and vital forces together. There are of course considerations that bear differentially on the existence of such forces, but I shall be proceding at a level where these are not significant.

Two arguments

My central suggestion will be that two rather different lines of evidence contributed to the demise of special forces. The first was an abstract argument based on theoretical physics, while the second was a more direct empirical argument based on physiological research. The first, abstract argument involves considerations to do with the conservation of energy, and was available from the time of Helmholtz onwards (even though it was not incontrovertible, and many were not persuaded). By contrast, the second, more direct argument does not follow from the conservation of energy alone, and indeed did not really gain force until the twentieth century.

At the end I shall argue that both arguments can be seen as contributing to the general modern acceptance of the completeness of physics. But the precise timing of this acceptance, and in particular the arrival of a general consensus in the second half of the twentieth century, seems to call for explanation in terms of the build-up of direct evidence for the second argument, rather than in terms of the more abstract argument which had been available since the middle of the nineteenth century.

Let me begin by presenting the two arguments in outline.

1 *The argument from fundamental forces*. The first argument is that all apparently special forces characteristically *reduce* to a small stock of

basic physical forces which conserve energy. Causes of macroscopic accelerations standardly turn out to be composed out of a few fundamental physical forces which operate throughout nature. So, while we ordinarily attribute certain physical effects to 'muscular forces', say, or indeed to 'mental causes', we should recognize that these causes, like all causes of physical effects, are ultimately composed of the few basic physical forces.

2 *The argument from physiology*. The second argument is simply that there is no direct evidence for vital or mental forces. Physiological research reveals no phenomena in living bodies that manifest such forces. All organic processes in living bodies seem to be fully accounted for by normal physical forces.

I take both of these to be empirically-based arguments, and both to have the same conclusion, namely, that there are no special mental or vital forces. But note that the evidential basis for the two arguments is quite different. The second argument appeals directly to the evidence uncovered by physiological research. It notes that observations made inside living bodies never reveal any accelerations that cannot be attributed to normal physical forces. The first argument, by contrast, appeals to the investigation of forces in general. It rests on evidence that many apparently different kinds of forces turn out to be composed of a few fundamental forces, and then applies this lesson to vital and mental forces in particular. So it need not appeal directly to any evidence about what goes on in living bodies. Instead it can infer the general conclusion inductively from the study of other forces, and then project it to the special case of mental and vital forces.

The argument from fundamental forces

Let me now explain the first argument more fully. I shall return to the second below. I take the materials for the first argument to have been available from the middle of the nineteenth century, and to relate to the reasoning which led up to the acceptance of the conservation of energy. It is true, as I have stressed, that the doctrine of the conservation of energy is itself consistent with the existence of special forces, as long as those forces are themselves conservative. At the same time, it seems to me that the thinking which supported the conservation of energy also weighed against special mental or vital forces.

At its simplest, my thought here is that the arguments behind the conservation of energy give inductive reason to suppose that all forces reduce to a small number of fundamental forces. We have already seen how Helmholtz's formulation of the conservation of energy hinged on the assumption that friction and other dissipative forces are non-fundamental forces, macroscopic manifestations of processes involving more fundamental conservative forces. For it is only if we see macroscopic forces like friction as reducing to fundamental conservative forces that we can uphold the universal conservation of energy. Now, this point can be viewed as providing inductive support for the general thesis that *all* apparently special forces will reduce to a small stock of fundamental forces. The special forces which have been quantitatively analysed, like friction, turn out to reduce to more fundamental conservative forces. So this provides inductive reason to conclude that any other apparently special forces, like muscular forces, or vital forces, or mental forces, will similarly reduce.[17]

This is of course not a knock-down argument. Vital or mental forces could themselves figure among the fundamental forces of nature, even if they are only generated in the special circumstances associated with life or sentience. But this position does not sit happily with a inductively based commitment to the universal conservation of energy. An insistence on the independent existence of *sui generis* special forces inside bodies threatens to remove the inductive reasons for believing in the conservation of energy in the first place. For there are no obvious grounds for expecting such *sui generis* special forces to be conservative.

After all, what argument was there, in 1850, say, for believing that forces operating inside bodies do not violate the conservation of energy? I am suggesting that the most persuasive argument hinged on the assumption that all forces operating in special circumstances reduce to a small stock of fundamental conservative forces. However, suppose now that it is explicitly specified that vital and mental forces do *not* reduce to other forces. Now we need independent evidence for supposing they are conservative, and it is not clear where it is to come from. In effect, then, positing *sui generis* vital or mental forces threatens to undermine the inductive grounds for upholding the conservation of energy in the first place. For it makes the assumption of their conservativeness an independent assumption, an assumption for which we lack any independent evidence.

I suspect that something like this line of thought lay behind Helmholtz's and his younger contemporaries' conviction that there were no special vital forces. Consider how Helmholtz argues in *Uber die Erhaltung die Kraft*. He takes pains to stress how it is specifically *central* forces independent of time and velocity which ensure the conservation of energy. This emphasis on central forces (by which Helmholtz meant forces which act along the line between the interacting particles) now seems dated. Nowadays, conservativeness is normally defined circularly, as a property of those forces which do no work round a closed orbit, and which are therefore the gradient of a scalar that depends only on position. This definition does not require a restriction to central forces. However, Helmholtz was in no position to adopt our circular definition of conservativeness. He was aiming to *persuade* his readers of the general conservation of energy, and so needed an argument. It would not have served simply to observe that energy is conserved by those forces which conserve energy. Helmholtz's actual claim was that energy is conserved by a wide range of known forces, namely, central forces. Still, this by itself does not show energy is conserved by all forces, *unless* all forces are central. Why should this be? Well, as above, the most plausible thought is surely that there is a small stock of basic central forces, and that all causes apparently peculiar to special circumstances are composed out of these.

To repeat, this is not a conclusive argument. Those thinkers who remained convinced, for whatever reasons, that there must be irreducible special forces inside living bodies, could still respect the universal conservation of energy, by maintaining that these extra forces must themselves operate conservatively. In support of this they could have offered the alternative inductive argument that, since all the *other* fundamental forces so far examined have turned out to be conservative, we should infer that any extra vital or mental fundamental forces will be conservative too.

I am not sure how far these alternative lines of inductive reasoning can be found explicitly laid out in the nineteenth-century debates. But they offer one possible explanation for the two different views on *sui generis* special forces which coexisted after the emergence of the conservation of energy. The thought that all apparently special forces reduce to a small stock of fundamental forces can account for the rejection of irreducible vital or mental forces by thinkers like Helmholtz and his young colleagues. Yet there were at least as many who wanted to maintain that vital and mental forces are *sui generis*, and they had the option of arguing that, even if these forces are fundamental and irreducible, the nature of other fundamental forces provides inductive reason to suppose these *sui generis* forces will be conservative in their own right.[18]

In connection with this latter school of thought, I have already mentioned Leibig and Müller, two eminent physiologists of the older generation, who continued to accept vital forces, even after the conservation of energy had won general acceptance. And Brian McLaughlin, in his excellent article on 'British Emergentism' (McLaughlin 1992), explains how the philosophers J. S. Mill and Alexander Bain went so far as to argue that the conservation of energy, and in particular the notion of potential energy, lends definite support to the possibility of non-physical forces. (The 'British Emergentists' discussed by McLaughlin were a philosophical movement committed precisely to non-physical causes of motion in my sense, causes which were not the vectorial 'resultants' of basic physical forces like gravity and impact, but which 'emerged' when matter arranged itself in special ways. The particular idea which attracted Mill and Bain was that these 'emergent forces' might be stored as unrealized potentials, ready to manifest themselves as a causes of motion only when the relevant special circumstances arose.)[19]

The argument from physiology

McLaughlin explains how British Emergentism continued to flourish into the twentieth century.[20] This highlights the question with which I began this paper. Given that thinkers continued to posit special mental and vital forces until well after the First World War, why has the idea of such forces now finally fallen into general disfavour?

Here I think we need to refer to the second line of argument against such forces, the argument from direct physiological evidence. We can view this second argument as operating against the background provided by the earlier argument from fundamental forces. The earlier argument suggested that at least most natural phenomena, if not all, can be explained by a few fundamental physical forces. This focused the issue of what kind of evidence would demonstrate the existence of extra mental or vital forces. For once we know which other forces exist, then we will know which anomalous accelerations would indicate the presence of special mental or vital forces. Against this background, the argument from physiology is then simply that detailed modern research has failed to uncover any such anomalous physical processes.

The relevant research dates mostly from the twentieth century. While important physiological research was carried out in the second half of the nineteenth century (see Coleman 1971), it did not penetrate to the level of forces operating inside bodies. At most it identified the chemical inputs and outputs to various parts of the body, and showed that animals are subject to general conservation principles. (See in particular Coleman 1971: 140–3, for Max Rubner's elaborate 1889 respiration calorimeter experiments showing that the energy emitted by a small dog exactly corresponds to that of the food it consumes.) Experiments of this kind, however, failed to provide compelling evidence against vital or mental forces. That normal chemicals are moved around, and that energy is conserved throughout, does not in the end rule out the possiblility that some accelerations within bodies are due to special vital or mental forces. It may still be that such forces are activated inside cells, but operate in such a way as to 'pay back' all the energy they 'borrow', and vice versa.[21]

In the first half of the twentieth century the situation changed, and by the 1950s it had become difficult, even for those who were not moved by the abstract argument from general reducibility, to continue to uphold special vital or mental forces. A great deal became known about biochemical and neurophysiological processes, especially at the level of the cell, and none of it gave any evidence for the existence of special forces not found elsewhere in nature.

During the first half of the century the catalyctic role and protein constitution of enzymes were recognized, basic biochemical cycles were identified, and the structure of proteins analysed, culminating in the discovery of DNA. In the same period, neurophysiological research mapped the body's neuronal network and analysed the electrical mechanisms responsible for neuronal activity. Together, these developments made it difficult to go on maintaining that special forces operate inside living bodies. If there were such forces, they could be expected to display some manifestation of their presence. But detailed physiological investigation failed to uncover evidence of anything except familiar physical forces.

In this way, the argument from physiology can be viewed as clinching the case for completeness of physics, against the background provided by the argument from fundamental forces. One virtue of this explanation in terms of two interrelated arguments is that it yields a natural explanation for the slow advance of the completeness of physics through the century from the 1850s to the 1950s. Suppose that we rank different thinkers through this period in terms of how much specifially physiological evidence was needed to persuade them of completeness, in addition to the abstract argument from fundamental forces. Helmholtz and his colleagues would be at one extreme, in deciding for completeness on the basis of the abstract argument alone, without any physiological evidence. In the middle would be those thinkers who waited for a while, but converted once initial physiological research in the first decades of this century gave no indication of any forces beyond fundamental forces found throughout nature. At the other end would be those who needed a great deal of negative physiological evidence before giving up on special forces. The existence of this spectrum would thus explain why there was a gradual build-up of support for the completeness of physics as the physiological evidence accumulated, culminating, I would contend, in a general scientific consensus by the 1950s.[22]

Conclusion

The problem I set myself at the beginning of this paper was to explain the rise of physicalist doctrines in the second half of this century. My argument has been that this is due to contemporary agreement on the completeness of physics. In the main body of this paper I have sought to show that this consensus is not just a fad, but a reflection of developments in empirical theory. Though it has not always been so, there is now good reason to believe the empirical thesis that all physical effects are due to physical causes. In particular, by the 1950s there was enough physiological evidence to persuade even those scientists who were unmoved by the abstract argument from fundamental forces.

The rise of physicalism among philosophers can be seen as a reflection of this development within science. Without the completeness of physics, there is no compelling reason to identify the mind with the brain. But once the completeness of physics became part of established science, scientifically informed philosophers realized that this crucial premise could be slotted into the various alternative versions of the causal argument for physicalism. There seems no reason to look any further to explain the widespread philosophical acceptance of physicalism since the 1950s.

Of course, as with all empirical matters, there is nothing certain here. There is no knock-down argument for the completeness of physics. You could in principle accept the rest of modern physical theory, and yet continue to insist on special mental forces, which operate in as yet undetected ways in the interstices of intelligent brains. And indeed there do exist bitter-enders of just this kind, who continue to hold out for special mental causes, even after another half-century of ever more detailed molecular biology has been added to the inductive evidence which initially created a scientific consensus on completeness in the 1950s. Perhaps it is this possibility which Stephen Clark has in mind when he doubts whether any empirical considerations can 'disprove' mind-body dualism. If so, there is no more I can do to persuade him of the completeness of physics. However, I see no virtue in philosophers refusing to accept a premise which, by any normal inductive standards, has been fully established by over a century of empirical research.[23]

Notes

1 Though see pp. 182–3 for some necessary qualifications.

2 Or, even more precisely, to accommodate quantum mechanical indeterminism: the *chances* of all physical occurrences are fully determined by a purely physical prior history. I shall ignore this qualification in nearly all that follows, since it would only complicate the issues unnecessarily.

3 Note, however, that while this is just a doctrine about physics, it does implicitly distinguish physics from other realms, since most other realms manifestly are not complete in this sense. The mental is not complete, for example, since there is no mental cause for the pain I feel when I sit on a drawing pin. Nor is the economic, since there is no economic cause for the economic costs occasioned by a hurricane. (This is why we do not find arguments aiming to

show that everything is mental, or economic, parallel to the completeness-based argument that everything is physical.)

4 In other wrtings, the relevance of the completeness of physics does not need to be excavated, since it lies on the surface. Thus see Feigl 1958, Oppenheim and Putnam 1958.

5 Equally: all chemical/biological/social occurrences have physical effects. The causal argument provides a schema that delivers physicalism for other special subjects as readily as for the mental. In the historical discussion that follows, various different special categories will be at issue at different points. But it will often be expositorily convenient to let the mental stand for the other cases, especially when addressing issues of argumentative structure rather than historical substance. The context should make it clear when the category of the mental is so being used.

6 However Sturgeon 1998 argues that an equivocation between a quantum-theoretical sense of 'physical' (in premise one) and an everyday sense (in premise two) invalidates the argument. This raises a number of interesting issues which I shall not be able to discuss here. But see Noordhof 1999 and Witmer forthcoming.

7 Of course, many philosophers are moved to pay this price because they cannot believe that *conscious* occurrences in particular can be identical with physical occurrences. I do not think that this is a good motivation. However, I do accept that physicalists owe some explanation of why conscious occurrences seem so very different from physical ones, if they are not. (See Papineau 1993 chapter 4; Papineau 1998).

8 Will we not have two causes anyway, namely (a) the role property with which we are now identifying the conscious property, and (b) the physical property which directly causes the behavioural result? We might in a sense have two 'causes', but they will not *overdetermine* the result, if the role property is present only in virtue of the physical property's presence. (Note that in this case the behavioural result would *not* still have occurred if the physical property had been absent, for then the role property would have been absent too; and similarly, if the role property had been absent in any particular case, so would the physical property have been absent.)

9 Conversely, those philosophers who do think that mathematical or moral facts have physical effects (in our brains, say) will come under pressure from the causal argument to identify them with physical facts.

10 Leibniz took it that all basic material particles are perfectly elastic, and that no kinetic energy is lost when they collide. He explained the apparent loss of kinetic energy when inelastic *macroscopic* bodies collide by positing increased motion in the microscopic parts of those bodies.Thus he explains, in the fifth paper of the *Leibniz–Clarke Correspondence* (Alexander 1956):

> The author objects, that two soft or un-elastic bodies meeting together, lose some of their force. I answer, no. 'Tis true, their wholes lose it with respect to their total motion; but their parts receive it, being shaken (internally) by the force of their concourse.

11 Cf. Papineau 1977.

12 Throughout the rest of this chapter I shall talk in terms of 'forces'. However, the issues will arise in just the same way if you regard forces as otiose, and instead think of the circumstances which 'cause forces' as themselves directly

causing the resulting accelerations. In that case, you will replace the question of whether there are 'mental forces' with the question of whether specifically mental initial conditions (conditions of sentience, or intentionality, or intelligence, depending on how you wish initially to pick out the mental) make a difference to accelerations, in the sense of entering as antecedents into special laws about accelerations which do not follow from other laws about accelerations. More simply, are there special accelerations in brains which are not predicted by other laws about accelerations? (Cf. McLaughlin 1992: 64–5.)

13 One barrier to the formulation of an energy conservation principle by early Newtonians was their lack of a notion of potential energy, the energy 'stored up' after a spring has been extended or compressed, or as two gravitating bodies move apart. Given this, there was no obvious sense in which they could view two gravitating bodies, for example, as conserving energy while they moved apart; after all, the sum of their kinetic energies would not be constant, but unequivocally decreasing. And even in the case of impact, where the notion of potential energy is not immediately needed, early Newtonians displayed no commitment to the conservation of (kinetic) energy. Most obviously, Newton and his followers were perfectly happy, unlike Leibniz, to allow unreduced inelastic collisions, in which both bodies lose kinetic energy without transmitting it to their internal parts. It is also worth remarking that there is nothing in Newton's Laws of Motion to rule out even 'superelastic' impacts, in which total kinetic energy increases. If two bodies with equal masses and equal but opposite speeds both rebounded after collision with double their speeds, for example, Newton's three Laws of Motion and the conservation of momentum would be respected. True, any such phenomenon would provide an obvious recipe for perpetual motion, but the point remains that Newton's Laws themselves do not rule it out. (It is also worth noting that perpetual motion was by no means universally rejected by seventeenth- and eighteenth-century physicists. Cf. Elkana 1974: 28–30.)

14 Here I am closely following Steigerwald 1998, chapter 2.

15 One model for this preservation was the kinetic theory of heat, which took the macroscopic kinetic energy apparently lost to be converted into internal kinetic energy at the microscopic level (cf. Leibniz's explanation for the apparent loss of kinetic energy in inelastic impact mentioned in note 10). But the abstract point at issue did not demand acceptance of the kinetic theory, since the lost kinetic energy could alternatively be viewed as being stored in the 'tensions' of whatever force might be associated with heat.

16 Helmholtz used the word '*Kraft*'. This is now standardly translated as 'force' rather than 'energy', but these two concepts were not clearly distinguished at the time, in either English or German. The general expectation at the time was that any conservation law would involve 'force' ('*Kraft*', '*vis*') where this was thought of as a directed quantity ('force of motion'), rather than as a scalar like energy. (Here again we see the dominance of the Newtonian tradition, whose only conserved quantity was the vectorial momentum.) One of Helmholtz's most important contributions was to make it clear that even within the Newtonian tradition of rational mechanics it is the scalar energy that is conserved, rather than any vectorial 'force'. Even so, the confusions persisted for some time, as shown, for example, by Faraday's 1857 paper 'On the conservation of force' (cf. Elkana 1974: 130–8).

17 This conservation-of-energy-based argument from fundamental forces raises some interesting questions in connection with quantum mechanics. (I am grateful to Barry Loewer for pressing these points on me.)

First, an initial query relates to my continued presentation of the issues in terms of forces. How does this fit in with modern quantum mechanics, which is normally formulated in terms of Hamiltonians rather than forces, that is, directly in energetic terms? But there is no substantial issue here, since the Hamiltonians themselves can be seen as depending on the relevant forces (cf. McLaughlin 1992: 54).

Second, on some interpretations, quantum systems do not always respect the conservation of energy. While energy is conserved in the 'Schrödinger evolution' of quantum systems, it is apparently violated by 'wave collapses'. Some, including myself, take this to argue against wave collapses. But, even if you do not go this way, it does not matter for this paper, since (a) the argument from fundamental forces to completeness will still have weight even if conservation is restricted to Schrödinger evolutions, and (b) completeness itself is consistent with the indeterminacy of collapse outcomes, since the chances of those outcomes are still fixed by prior physical forces alone (cf. note 2).

Third, on some, but not all, collapse interpretations, *sui generis* factors do seem to fix whether a collapse occurs or not (even though the subsequent chances of the various possible outcomes then still depends entirely on the prior physical forces). I am thinking here of interpretations which say that collapses occur when physical systems interact with consciousness (or indeed which say that collapses occur when there are 'measurements', or 'macroscopic interactions', and then refuse to offer any physical reductions of these terms). On these interpretations, the completeness of physics is indeed violated, since collapses do not follow from more basic physical laws but depend on 'emergent' causes. It would seem an odd victory for non-specialists, however, if the sole locus of *sui generis* mental action were quantum wave collapses.

18 I have the impression that scientifically-informed late nineteenth-century philosophers were not particularly exercised by our issue of whether or not there are special vital or mental forces. Understandably enough, they were far more interested in the determinism which, as I have pointed out, is required by the conservation of energy even if we admit special mental forces (cf. Tyndall 1898[1877]).

19 Indeed this line of thought seems to have become extremely popular in the late nineteenth century. The idea that the brain is a repository of 'nervous energy', which gets channelled in various ways, and is then released in action, is a commonplace of Victorian thinkers from Darwin to Freud.

20 Not all emergentists were as sophisticated as Mill and Bain. In *Mind and its Place in Nature* (1923): 103–9, C. D. Broad addresses the issue of whether independent mental causation would violate the conservation of energy. But instead of simply claiming that any mental force would operate conservatively, he insists that the principle of the conservation of energy does not explain all motions, even in physical systems, and so leaves room for other causes. He draws an analogy with a pendulum on a string, where he says that the 'pull of the string' is a cause which operates independently of any flows of energy, and he suggests that the mind might operate as a similar cause. While it is not entirely clear how Broad intends this analogy to be read, it is difficult to avoid the impression that he has mastered the letter of the principle of the conservation of energy without grasping the wider physical theory in which it is embedded.

21 Indeed, and somewhat paradoxically, this species of 'bookkeeping' experiment may even have weighed in *favour* of postulating *sui generis* vital forces. This

is because these experiments offer a counter to the argument from fundamental forces. That argument, remember, hinged on the claim that there is no direct inductive reason to suppose that any *sui generis* vital forces are conservative, if it is denied that they reduce to more fundamental forces. But experiments like Rubner's do offer just such direct inductive reason, in that they show that any special forces operating inside bodies must always 'pay back' just as much energy as they 'borrow', even if they do not reduce to more fundamental forces. (I owe this point to Keith Hossack.)

22 McLaughlin (1992: 89) attributes the end of British Emergentism, and therewith the rise of contemporary physicalism, to the 1920s quantum-mechanical reduction of chemical forces to general physical forces between sub-atomic components. But it seems unlikely that this could have been decisive. After all, why should anybody be persuaded against special mental causes just because of the reduction of *chemistry* to physics? (Why should it matter to the existence of *sui generis* mental forces how many independent forces there are at the level of atoms?) At most the reduction of chemistry to physics would have added weight to the argument from fundamental forces, by showing that yet another special force reduces to more basic forces. But it was irrelevant to the argument which I claim swayed thinkers in the twentieth century, the argument from physiology.

23 I should greatly like to thank Barry Loewer, Keith Hossack, James Ladyman, Joan Steigerwald, Scott Sturgeon and David Spurrett for comments on drafts of this paper.

Bibliography

Alexander, H. (ed.) (1956) *The Leibniz-Clarke Correspondence*, Manchester: Manchester University Press.

Armstrong, D. (1968) *A Materialist Theory of the Mind*, London: Routledge and Kegan Paul.

Ayer, A. J. (1969) 'Man as a Subject for Science', in P. Laslett and W. Runciman (eds) *Philosophy, Politics and Society 3rd Series*, Oxford: Blackwell.

Broad, C. D. (1923) *Mind and its Place in Nature*, London: Routledge and Kegan Paul.

Clarke, S. (1996) 'Review of C. Taliaferro, *Consciousness and the Mind of God*', *Times Literary Supplement* 23 February.

Coleman, W. (1971) *Biology in the Nineteenth Century*, New York: Wiley.

Crane, T. (1995) 'The mental causation debate', *Aristotelian Society Supplementary Volume 69.*

Davidson, D. (1963) 'Actions, reasons and causes', *Journal of Philosophy* 60.

Davidson, D. (1970) 'Mental events', in L. Foster and J. Swanson (eds), *Experience and Theory*, London: Duckworth, reprinted in Davidson (1980).

Davidson, D. (1980) *Essays on Actions and Events*, Oxford: Clarendon Press.

Dray, W. (1957) *Laws and Explanation in History*, Oxford: Oxford University Press.

Elkana, Y. (1974) *The Discovery of the Conservation of Energy*, London: Hutchinson.

Feigl, H. (1958) 'The "mental" and the "physical"', in H. Feigl, M. Scriven and G. Maxwell (eds) *Minnesota Studies in the Philsohy of Science vol II*, Minneapolis:

University of Minnesota Press.
Faraday, M. (1857) 'On the conservation of force', *Procedings of the Royal Society*.
Goldstein, H. (1964) *Classical Mechanics*, 2nd edn, Reading, Mass.: Addison-Wesley.
Haller, A. von (1936 [1751]) 'A dissertation on the sensible and irritable parts of animals', trans, O. Temkin, *Bulletin of the History of Medicine* 4.
Helmholtz, H. von (1847) *Uber die Erhaltung die Kraft*, Berlin.
Hempel, C. (1942) 'The function of general laws in history', *Journal of Philosophy* 39.
Leibniz, G. (1898 [1696]) *The Monadology*, trans. R. Latta, Oxford.
Lewis, D. (1966) 'An argument for the identity theory', *Journal of Philosophy* 63.
Lewis, D. (1980) 'Mad pain and Martian pain', in N. Block (ed.) *Readings in the Philosophy of Psychology vol 1*, Cambridge, Mass.: Harvard University Press.
McLaughlin, B. (1992) 'The rise and fall of British emergentism', in A. Beckerman, H. Flohr and J. Kim (eds) *Emergence or Reduction*, New York: De Gruyter.
Mellor, D. H. (1995) *The Facts of Causation*, London: Routledge.
Nagel, E. (1961) *The Structure of Science*, London: Routledge and Kegan Paul.
Newton, I. (1960 [1686]) *Mathematical Principles of Natural Philosophy*, ed. F. Cajori, Berkeley: University of California Press.
Newton, I. (1952 [1704]) *Opticks*, New York: Dover.
Noordhof, P. (1999) 'The overdetermination argument versus the cause-and-essence principle – no contest', *Mind* 108.
Papineau, D. (1977), 'The *vis viva* controversy: do meanings matter?' *Studies in History and Philosophy of Science* 8.
Papineau, D. (1993) *Philosophical Naturalism*, Oxford: Blackwell.
Papineau, D. (1998) 'Mind the gap', in J. Tomberlin (ed.) *Philosophical Perspectives* 12.
Papineau, D. and Spurrett, D. (1999) 'A note on the completeness of physics', *Analysis* 59.
Place, U. (1956) 'Is consciousness a brain process?', *British Journal of Psychology* 47.
Putnam, H. (1960) 'Minds and machines', in S. Hook (ed.), *Dimensions of Mind*, New York: New York University Press.
Oppenheim, P. and Putnam, H. (1958) 'Unity of science as a working hypothesis', in H. Feigl, M. Scriven and G. Maxwell (eds) *Minnesota Studies in the Philsohy of Science vol II*, Minneapolis: University of Minnesota Press.
Segal, G and Sober, E. (1991) 'The causal efficacy of content,' *Philosophical Studies* 63.
Smart, J. (1959) 'Sensations and brain orocesses', *Philosophical Review* 68.
Steigerwald, J. (1998) *Lebenskraft in Reflection*, Ph.D. dissertation, University of London.
Sturgeon, S. (1998) 'Physicalism and overdetermination', *Mind* 110.
Tyndall, J. (1898 [1877]) 'Science and Man', in his *Fragments of Science, vol 2*, New York.
Whytt, R. (1755) *Physiological Essays*, Edinburgh.
Winch, P. (1958) *The Idea of a Social Science*, London: Routledge and Kegan Paul.
Witmer, G. (forthcoming) 'Causal overdetermination and the macro-micro gap'.
Woolhouse, R. (1985) 'Leibniz's reaction to Cartesian interaction', *Proceedings of the Aristotelian Society* 86.

11 Against the completability of science

Nancy Cartwright

Physics and its *ceteris paribus* laws

One of the most lofty ambitions that science can entertain is *completeness*, the hope to account for everything that happens. Usually the vision of completeness is combined with some kind or another of reductionism, where the science of choice – the science that will account for everything – is physics. I shall argue here that we have good reasons to think this grand goal is not achievable, not even 'in principle'.

It might well be that the plausibility of this doctrine depends on what is demanded in order to say that a science 'accounts for' something. I shall take it that a science accounts for a feature of an event when the laws or theories of the science can provide a regularity that subsumes the occurrence of that feature. The regularity can involve features that are very specific or ones that are very general, ones that can be observed or ones that are defined primarily through the roles they play in theory; it can be either factual or counterfactual; it can be either deterministic or probabilistic; and it can be either associationist or causal.[1]

This may look to push the problem back to a question of what the criteria are for a regularity to subsume some feature of an event. We clearly want to include any cases where the feature can be paired with some initial or boundary conditions to provide an instance of the regularity. But what else might count? In particular, given the popular doctrines of the last twenty years, we may want to know whether a feature is subsumed by the regularities given in physics if those regularities couple with initial or boundary conditions to fix (at least the probability for) the occurrence of a 'physics' state on which the feature supervenes.

As it turns out, for my arguments here we will not need to settle these matters. For supervenience, along with more classical versions of theory reduction, all advocate what we can think of as *downwise* reduction: in some sense or another, everything that is not in the domain of physics is to be reduced to what is. I shall argue instead that we do not even have good reason to believe in *crosswise* reduction: physics cannot account for everything that is in its domain.

The reason for this is that the regularity claims that can be provided by our theories in physics are all about a certain particular kind of *ceteris paribus* regularity. Physics has achieved amazing success at isolating domains of features for which there is a kind of closure among their causes: these features have a set of causes which are sufficient to fix at least their probabilities *so long as no causes of these features occur other than ones in the set*.

I have come to believe in this claim by looking at a large number of cases of how physics works in the world. Many philosophers of physics proceed in a quite opposite way. They wish to understand what physics theories say. In my own field of philosophy of quantum mechanics, for example, we wish to know what quantum mechanics says about non-local causation or about well-defined states for macro-objects, or we wish to know what renormalisation really means or whether symmetries have replaced laws.

To answer these questions philosophers of physics tend to engage in detailed investigations of the mathematical structure of the theory in question. I think this is a mistake. What a theory says is exactly what it needs to say in order to account for empirical phenomena in the way that it does. I borrow an argument from the realist debate. On empiricist cannons we are justified in accepting a theory precisely because of its empirical successes. But then the content of the theory that we are entitled to accept is just the content necessary to produce those empirical successes. So, to see what a theory says, it is essential to look in detail at how it is put to use.

My conclusion from looking at a large number of cases of how theories in physics are used to treat real situations in the world, both in testing the theories and in their impressive technological applications, is that it is always *ceteris paribus* regularities that come into play.[2] All the cases I have looked at have just the characteristic I point to: they are either especially engineered or especially chosen *to include only those causes that occur in the preferred set* of the theory. They are, moreover, always arranged in a very special way: a way that the theory knows how to describe and to predict from. That is not surprising where *ceteris paribus* laws are involved, since we can neither test laws of this kind nor apply them until we are sure the *ceteris paribus* conditions are satisfied. The point is that these are the kinds of cases that give us our most powerful reasons for accepting our theories in physics. And the laws they give us reason to accept are all *ceteris paribus* laws.

Besides brute fact observation, there are a number of independent considerations that allow us to make sense of this claim about *ceteris paribus* laws and to see how and why it might be true. Together these considerations show how we can shift our image of laws to admit the *ceteris paribus* character we can observe them to have without losing either the intelligibility that we know nature possesses or the possibility of a sensible reconstruction of scientific knowledge. Specifically, I argue for the following:

- The constraint on the scope of theories that I describe follows naturally from our standard account of what scientific theories are, although the point becomes somewhat obscured when we give a semantic rather than a syntactic version of this account.
- In general, regularities occur only in very special kinds of circumstances arranged in very special ways. To highlight many of the special features of these kinds of circumstances I call them *nomological machines*. A nomological machine is a fixed (enough) arrangement of components or factors with stable (enough) capacities that in the right sort of stable (enough) environment will, with repeated operation, give rise to the kind of regular behaviour we describe in a scientific law.

This is not a characterization that will satisfy a Humean. This is especially true since it is part of my overall point about the structure of scientific theory that the basic principles that help us understand how explanatory, counterfactual-supporting regularities arise are not claims of universal or probabilistic association, even if we put a modal operator in front. They are, rather, claims about capacities.

Crudely the connection between nomological machines and *ceteris paribus* laws is straightforward. We are interested in machines that are structured in such a way that the regular behaviour they give rise to is predictable from theory, since these are the regularities that we appeal to in theoretical explanation and in forecasting. In order for the theory to make these predictions, both the factors and the arrangement in which they occur must be ones the theory knows how to treat. And the machine must be effectively 'shielded' so that nothing interferes that the theory can not deal with – then it will be in the 'right kind of stable environment'.

The counterfactual-supporting regularities that we prize hold, then, only relative to the successful operation of a nomological machine, and all the conditions necessary to its design and operation should appear in a large "*ceteris paribus*" clause in front of the regularity claim. When the behaviour of the machine is predictable from a theory, a number of these conditions can be moved into the antecedent of a proper theoretical claim. These are the ones that describe the components and their arrangement. But in general the conditions that describe the shielding cannot. These are the factors I have been referring to here, the ones that make our explanatory claims in physics irredeemable *ceteris paribus*.

It supports my claims about nomological machines in physics to notice that they are not peculiar to this discipline. Rather, our best understood regularities seem to arise from nomological machines across a variety of different areas that I have looked at, both in the natural and in the social sciences.

- There is a perfectly acceptable alternative to the image of a world neatly ordered under law, whether a single set of related laws or a federation of laws from different sciences. Indeed, the alternative image is far truer to our experiences of the world than is the image suggested by various theses of scientific completeness, of reductions of various sorts, and of claims that everything supervenes on one special kind of base, even should we allow global supervenience.

This is an image of a dappled world, a world rich in different things, with different natures, behaving in different ways. The laws that describe this world are a patchwork, not a pyramid. They do not take after the simple, elegant and abstract structure of a system of axioms and theorems. Rather they look like – and steadfastly stick to looking like – science as we know it: apportioned into disciplines, apparently arbitrarily grown up; governing different sets of properties at different levels of abstraction; pockets of great precision; large parcels of qualitative maxims resisting precise formulation; erratic overlaps; here and there, once in a while, corners that line up, but mostly jagged edges; and always the cover of law just loosely attached to the jumbled world of material things.

The evidence suggests that, for all we know, most of what occurs in nature occurs by hap, subject to no law at all. What happens is more like the outcome of negotiation between domains than like the logical consequence of a system of order. Within this image of a world not uniformly governed by the universal rule of law, we can make sense of the great pockets of regularity, both precise and imprecise, that we see or that we construct – the seasons, the planetary motions, the human body, various ecosystems, bicycles, computers, financial markets and the like – with the idea of the nomological machine. The dappled world is what, for the most part, comes naturally: regimented behaviour results from good engineering.

I develop these points in various ways in *The Dappled World: A Study of the Boundaries of Science* (Cartwright 1999). Here I shall focus on the first point since it provides the most immediately relevant argument against the completability of physics.

Causes that cannot be codified in theory

At the outset I should make some remarks in defence of the general background assumption that the idea of a cause which cannot be codified into a theory makes sense. After all, the standard objection to *ceteris paribus* laws is that there is no difference between a *ceteris paribus* regularity and no regularity at all. The consequent occurs as predicted from the antecedents unless . . . Unless what? Unless something occurs other than the designated antecedents? But something else always occurs.

I have not formulated the claim about closure in this way, though, but rather: the effect occurs as predicted from the designated causes unless some cause occurs not in the designated set. This formulation relies on the assumption that there are *sui generis* singular causal facts. It supposes that on any given occasion there is a fact of the matter about whether a factor of the wrong kind – a factor of a kind not in the designated set – operated to influence the effect or not.

In defense of this assumption I rely on a large philosophical literature. This includes arguments of my own in *Nature's Capacities and their Measurement* (Cartwright 1989) from both an ontological and an epistemological point of view. On the ontological side, I argue that we must reconstruct causal laws as claims about what kinds of singular causal facts will reliably occur. In particular they cannot be taken as any kind of claim about (probabilistic) association, nor about abstract relations between universals. Epistemically, we need to know singular causal facts in order to test claims of any other kind, including claims about general laws, just as we need to know claims about general laws in order to test singular causal claims.

It is, however, not necessary to make the kinds of strong assumptions about singular causal facts that I make in order to defend the proposed reading of *ceteris paribus* laws. We may instead take the weaker view that what makes a factor a causal factor is that it falls under some kind of causal law, so long as we do not demand that all laws express precise quantitative relationships like the equations of physics. All that is required to support the claim that the laws of physics do not dictate regularities that hold *tout court* is the assumption that the external causes not covered in the theory do not combine in any way acceptable in a law of physics with those that are included in the theory. They may, for instance, be purely qualitative in their influence, or, if quantitative, fail to compose in any systematic way with the other causes.

John Stuart Mill claimed that just this latter is true of causes in chemistry; though he believed, happily, it is not true of causes in economics. Mill also took the standard view that it would not be true of causes in mechanics: all causes of motion are forces and all forces add vectorially. That, I believe, is because he did not think seriously about all the qualitative factors that we know affect the motions of objects which we are not able to regiment under the concept of force: causes like the wind or the mischievous child who grabs the billiard ball off the table at the moment of collision. I shall discuss this kind of case in the next section.

A striking example

My claim about *ceteris paribus* laws supposes that the effects which physics studies have causes that do not fall within physics: causes that do not have the right kind of systematic relations with the effects they can influence to be included in a precise, predictive theory along with the kinds of causes physics familiarly treats. I repeat here the example that I gave in a paper for the Aristotelian Society, since it illustrates the point dramatically using the very familiar example of Newton's second law, $F_t = ma$ (I put the "t" in the subscript to make clear that we are talking about the *total* force) (Cartwright 1994).

Suppose we drop a small compact wooden sphere from a given height in a vacuum. The force of gravity is the total force operating on it, and nothing else affects its motion. In this case the motion that results will be an instance of the regular association described in Newton's law. Consider instead that we drop a floppier object in far less controlled circumstances. Otto Neurath has a nice example; my conclusions about it are much like his.

> In some cases a physicist is a worse prophet than a [behaviourist psychologist], as when he is supposed to specify where in St. Stephen's Square a thousand dollar bill swept away by the wind will land, whereas a [behaviourist] can specify the result of a conditioning experiment rather accurately.
>
> (Neurath 1933: 13)

The wind is not an instance of the force of gravity. Nor is it, on the face of it, an instance of any other kind of force: it does not look like *any* of our standard models of forces. We can try to get it to do so. For instance we can think of breaking it up into thousands of little nuggets of air each *colliding with* the bill. If we have sufficient reason for believing that there is an acceptable model like this countenanced by the theory, a model that would be both predictively accurate and descriptively true, then we have sufficient reason to think the wind is a cause that is represented among those that physics can treat.

I challenge our reasons for thinking we can do so. As Kelvin argued, mechanics can give fantastic predictions when it treats what is inextendable, unbendable and stiff: things that resemble compact masses, rigid rods, and springs. But it has never been able to simulate much of the soft, continuous, elastic and friction-full world around us. Kelvin argued exactly the opposite. He expected he could treat all motions as motions in a continuous flexible medium. But when it comes to treating real systems, Kelvin's takeover attempt fares no better than the Newtonian one.

What looks to be the case is that our spectacular successes come from using Newtonian particle mechanics where the causes of motion look like Newtonian models, and Kelvin-like theories where they look like Kelvin's view of the world; and in the cases where we do have successes where the causes are mixed, that is by a lot of *ad hoc* patching that does not appeal to any theoretically-grounded regularities. Any particular case may be debatable: can we bring it under this theory or that theory or any theory? But the reasons, I take it we should agree, must be concrete empirical ones; a general confidence in the completability of some one or another theory of physics is not a good enough argument.

Bridge principles and the scope of theory

Perhaps the reasons we are inclined to give on behalf of the complete coverage of our favored theory for its special domain of effects are broadly inductive. One of the great points about Newton's rules of reasoning after all is that the same laws govern masses wheresoever they appear, in particular independent of whether they are terrestrial masses or celestial ones. And this assumption has been well borne out in hundreds of thousands of cases.

This raises one of the most central questions we face in philosophy of science: what should be the bounds on our inductions? Theory of confirmation is notoriously difficult, and all the more so once we give up the requirement that theories be reconstructed in special formal languages. I should like to appeal to a crude intuitive principle: when we can recognize a clear boundary within which all our successful cases have been located and, moreover, we can offer good reasons why that boundary might well be relevant, then failing compelling reason to the contrary, we should not extend our inductions beyond that boundary.

For a large number of theories in physics that I have looked at, I think we have such a clear boundary: the empirical successes of the theory are all for cases that fit the theory's interpretive models, or better, that fit some arrangement licensed by the theory of its interpretive models. This, I claim, is true for classical mechanics, classical electromagnetic theory, classical and quantum statistical mechanics, quantum field theory, quantum electrodynamics, and condensed matter physics. It may be true for other theories as well; these are just the ones that play a central role in the cases I have looked at in some detail.

Let me explain about interpretive models. C. G. Hempel and his contemporaries taught us that the basic principles of theory are of two kinds, *internal principles*, which show relations among theoretical concepts, and *bridge principles*, which link the theoretical concepts with other more concrete or more readily measurable or more well understood concepts. This important point still holds, though it is more cumbersome to express in the currently fashionable semantic accounts of theory. The models will contain both those properties that we label 'theoretical' and those we label 'observational', and only those models that exhibit the right relations among these will count as models of the theory.

Part of the reason that the distinction is not made much of nowadays is that it is commonly thought to have neither the epistemological nor the semantic importance that we used to attribute to it. For many, observation is 'theory-laden' in a way that denies it any epistemic privilege; nor does it provide the base from which the meanings of all other descriptive terms are built. I shall not discuss either of these points here for neither matters to my argument.

There is in addition a third role that bridge principles play that I shall focus on. Bridge principles provide the sets of permissible *concrete* descriptions that cash out the *abstract* concepts of theories in physics.[3] All the theories that I have mentioned use abstract concepts. 'Abstract' has a great variety of different senses. I mean it here in a very specific sense I borrow from the German Enlightenment thinker and playwright, Gotthold Ephraim Lessing (Lessing 1759 [1967]).

The account of abstraction that I borrow from Lessing to describe how contemporary physics theories work provides us with two necessary conditions. First, a concept that is abstract relative to another more concrete set of descriptions never applies unless one of the more concrete descriptions also applies. These are the descriptions that can be used to 'concretize' the abstract description on any given occasion. Second, satisfying the associated concrete description that applies on a particular occasion is what satisfying the abstract description consists of on that occasion.

Writing this paper is what my working right now consists of; being located at a distance r from another charge q_2 is what it consists in for a particle of charge q_1 to be a subject to the Coulomb force $q_1q_2/4\pi\varepsilon_o r^2$ in the usual cases when that force function applies. To say that working consists of a specific activity described with the relevant set of more concrete concepts on any given occasion implies at least that no further description using those concepts is required for it to be true that 'working' applies on that occasion, and similarly for other abstract/concrete pairs.[4] Surely the notion is richer than this, though I do not yet have anything I can confidently add.

A number of the quantities we use in our modern theories in physics are abstract in just this way. This includes the force and energy functions of classical mechanics, the electric and magnetic field vectors, the quantum Hamiltonian, and quantum fields. But a number of equally central concepts are not abstract, such as the concepts of *mass*, *charge*, *acceleration*, and the *quantum state*, along with most physical constants and parameters, like the gravitational constant or the conductivity of an electrical medium.

For example, whenever a force function correctly describes a situation, there is an additional more concrete description that is what the obtaining of that force function consists in, as with the Coulomb force example in the previous paragraph. But this is not the case when we assign a given charge to a body. Of course a large number of other things may always be true of a body when it is charged, but they do not constitute what it is to be charged on that occasion.

If we look to the appropriate theory, for each of these abstract concepts we find bridge principles that tell us what are the set of appropriate descriptions on which the abstract feature can piggy-back. Following conventional usage, I call the more concrete descriptions picked out by the theory the *interpretive* models of the theory, though they would more perspicuously in this role be called *concretizing* models.

My point can be illustrated by looking at the table of contents of a typical mechanics text, for example, the classical text by Robert Lindsay (1961). A major part of a book like this is aimed at teaching you what interpretive models can serve as the correlates of what force functions. The text begins with a simple chapter of introduction, 'The elemental concepts of mechanics'. Already the second chapter starts to offer simple arrangements and their force functions and the third continues with more complicated models. The fourth chapter introduces energy and immediately turns to an account of what energy functions can be matched to what situations: 'Energy relations in a central force field', 'Inverse square field', 'Electron energies in the Bohr atom.' The same pattern is followed in the discussion of equilibrium (*e.g.* 'Equilibrium of a particle. Simple cases . . . A system of particles . . . Equilibrium of a flexible spring') and similarly throughout the text.

We see the same thing in texts in electromagnetism, in optics, in quantum mechanics, and so on; in advanced texts as well as in elementary texts; and in actual usage. Bridge principles play an essential role in these theories and certain concepts – *force*, *energy*, *equilibrium* and the like – are never properly introduced into the treatment of a situation without them. I say 'properly' because of course they often are introduced. For instance we frequently add 'phenomenological' terms to a force function or to a quantum Hamiltonian to make up for the difference between the effects that have been modelled and those that are measured. But they are marked out in a special way – as by the label 'phenomenological' – to highlight the fact that they are not introduced in a principled way.

The point is: this is how abstract concepts are assigned in a proper theoretical treatment; and in particular it is how they are assigned in *just those derivations that we take to be the best cases where predictive success argues for the truth of the theory*. When the abstract concepts do not piggy-back on the specific concrete features ascribed to a situation – that is, when there is no bridge principle that licenses their application to the situation – then their introduction is *ad hoc* and the power of the derived prediction to confirm the theory is much reduced. For the cases that give us really good reasons to believe in the truth of a theory, we need descriptions that are assigned in a principled way; and for the theories I have mentioned as they are practiced that means ones that are licensed by principles of the theory: by bridge principles.

Now when we think about the range and limits of theory there is something to be noticed. Bridge principles are just like internal principles in one respect: there are just a handful of them. And that is in keeping with the point of abstract theory as it is described by empiricists and rationalists alike. We aim to cover as wide a range as we can with as few principles as possible. But the fewer the bridge principles, the fewer the concrete models we have available to describe the world. Since the theory can be applied only when its concrete interpretive models fit, the range of the theory will be severely restricted, even though the predictions within that range may be enormously precise and impressively accurate.

So, how far should our inductions from our empirical successes extend for the theories I have been talking about? The conclusion to be drawn from these observations about bridge principles and abstract theory is that our inductions should not go beyond the range where our interpretive models fit.

The case of superconductivity

Let us look at a specific example. We are invited to believe in the truth of our favourite explanatory theories because of their precision and their empirical successes. The Bardeen–Cooper–Schrieffer (BCS) account of superconductivity must be a paradigmatic case (Bardeen, Cooper and Schrieffer 1957: 1175). We build real operating finely-tuned superconducting devices using the Ginsburg–Landau equations. And (with appropriate corrections) we know that the Ginsburg–Landau equations can be derived from quantum mechanics or quantum field theory using the BCS model. So every time a SQUID (Superconducting Quantum Interference Device) detects a magnetic fluctuation we have reason to believe in quantum theory.

But the Hamiltonian used in the quantum equations of motion, like the classical force function, is abstract: we apply it to a situation only when that situation is deemed to satisfy certain other more concrete descriptions. These are the descriptions provided by the interpretive models of quantum mechanics. Albert Messiah's text *Quantum Mechanics* provides four basic interpretive models: the central potential, scattering, the Coulomb interaction and the harmonic oscillator, to which we should add the kinetic energy, which is taken for granted in his text (Messiah 1961). The quantum bridge principles give the corresponding Hamiltonians for each of these concrete interpretive models.

So far I have mentioned four basic bridge principles from Messiah. We may expect more to be added as we move from fundamental quantum theory to more specific theories for specific topics. But, in keeping with my remark at the end of the last section, we should not expect the set of bridge principles to be greatly expanded. The BCS account of superconductivity is a good example. I chose this case to study because it was one I knew something about from my work on SQUIDs at Stanford and from our research project on modeling at LSE. It turns out to be a startling confirmation of my point.

The fundamental Hamiltonian of the BCS paper uses only the basic models I have already described plus just one that is new: the kinetic energy of moving particles, the harmonic oscillator, the Coulomb interaction, scattering between electrons with states of well-defined momentum, and then in addition the 'Bloch' Hamiltonian for particles in a periodic potential. And this last is itself closely related to the central potential, which is already among the basic models. Superconductivity is a quantum phenomenon precisely because superconducting materials (at least low temperature, 'type-I' materials) can be represented by the special models that quantum theory supplies.

There is no logical necessity for quantum mechanics to work in this way, of course. The Hamiltonian might, for instance, have resembled the quantum state or charge or mass: it might have functioned as a concrete not an abstract concept. But this alternative kind of theory is not the theory we have. In particular it is not the theory whose amazing empirical successes provide us with powerful reasons for accepting it. In the theory we do have reason to accept and to use, the Hamiltonian is abstract. So the tie of the theory to the world is constrained by the interpretive models we have successfully developed for the Hamiltonian.

How much of the world altogether can be represented by these models is an open question. Not much, as the world presents itself, looks on the face of it like harmonic oscillators and Coulomb interactions between separated chunks of charge. Superconductivity is a case where a highly successful representation can be constructed from just the models quantum theory has to offer. My point is that with each new case it is an empirical question whether these models, or models from some other theory, or no models from any theory at all will fit. Quantum theory will apply to phenomena that these models can represent, and nothing in the theory, nor anything else we know about the structure of matter, tells us whether they can be forced to fit in a new case where they do not at first appearance do so.

Conclusion

I claim that the regularities that follow from a good many of our most highly prized theories in physics are *ceteris paribus* regularities: they hold only so long as all the causes that operate are causes described within the theory. This is a trivial and uninteresting claim – and a claim that clearly does not argue against the completability of physics – unless it is additionally supposed that there may well be causes of its targeted effects that the theory cannot describe. So, how far does the descriptive capacity of our theories stretch?

I have argued that we have good reason to be suspicious of the unlimited stretch of our theories because many of their central concepts are abstract: they do not correctly describe a situation unless a more concrete description obtains as well. Our theories in physics aim to lay out what systematic relations there are among the features they study. When it comes to the relations between abstract descriptions and the more concrete ones on which they piggyback, it is the bridge principles that do this job. Bridge principles provide us with very specific sets of descriptions – we usually call them 'interpretive' models – which provide concrete correlates of the abstract terms. In the context of this discussion my question then becomes: can the interpretive models of our theories correctly model all the causes of the effects in the domains of those theories?

The fact that a description is abstract is neither necessary nor sufficient for it to be limited in its application in the way in which I have claimed some abstract descriptions in physics are. Consider necessity first. To have a specific mass or a specific charge or a specific acceleration are all very concrete features, and none of these seem to be constrained to any special kinds of situations. Any system of a certain kind – an electron for instance – will have a mass or a charge or an acceleration regardless of the situation it is in. At least it looks to me as if that is what our successful uses suggest.

The quantum state also provides a very concrete description, as with *acceleration* and *charge*. But it is, at least as Nobel laureate Willis Lamb argues, very limited in its domain of application. It takes very specific kinds of preparations to produce any quantum state at all on his view, and the kinds of preparations will determine the range of states available. Clearly the same kind of thing can happen for any determinable feature whose causes have a fixed range of application. The determinates that fall under it will obtain only in those special circumstances where the causes occur. They can nevertheless be maximally concrete relative to descriptions of those circumstances.

Sufficiency is the interesting question here. What is sufficient is clearly that the abstract descriptions piggyback on more concrete descriptions that are themselves limited in their range of application. The theories I have been discussing all apply only as far as their interpretive models can stretch. And by the nature of how models do – and should – get constructed, the kinds of models available are very precise in their form and very limited in number. So they fit readily on to only very special bits of the world around us. Sticking to Messiah's catalogue of bridge principles as an example, that means that quantum theory extends to all and only those situations that can be represented as composed of central potentials, scattering events, Coulomb interactions and harmonic oscillators.

As I pointed out in the last section, not much of the world on the face of it looks like that. How things look does not of course entirely settle the matter. But it must be taken as strong *prima facie* evidence. In the face of it we need very good evidence to the contrary, and it is important that it be evidence – good empirical evidence – not metaphysics nor wishful thinking nor the substitution of an aesthetic ideal for sound methodology.

The Mechanical Philosophy was already powerful in the 1660s in the early Royal Society, and it remained so for over 150 years. We tried to reduce everything to the features studied in mechanics: electricity, magnetism, birth, development, the motion of the blood, chemistry, cooking . . . everything that behaved in any kind of systematic way. There were successes. But there were immeasurably more failures. And the failures were costly, both directly and in lost opportunities for developing other approaches. At the end of the nineteenth century we saw a reverse movement, this time to reduce all of physics and everything that follows from it to electromagnetics. This programme was equally unsuccessful at a general takeover.

Nothing in what I have said stops us from hoping that our favoured theory – or some as yet undreamt-of theory – is completable. But it gives us good reason to think it may not be. A good many philosophers have strong views about warrantability and belief: we should not believe what we do not have good reason for. Whatever is the case about belief, I want to urge that we take strong views about action. Where either hope for or belief in the completability of physics is going to affect what we do – for instance in giving one research programme an edge over another – we had better have sufficiently strong reasons for our choices, where the reasons had better be based on empirical evidence and empirical success. And when we do not have evidence enough, we had better hedge our bets and not resort to metaphysics, neither the metaphysics of the rule of law nor that of the dappled world.[5]

Notes

1 On the reading I prefer, causal laws make claims about what kinds of singular casual facts reliably occur in a specific kind of population, where 'reliably' has to be cashed out in some way that permits the claim to bemerely probabilistic.

2 The detailed studies I have done have concentrated on experiments and applications involving primarily classical mechanics, classic electromagnetic theory, classical quantum mechanics, quantum field theory and condensed matter theories, especially in the construction of lasers and superconducting devices.

3 For accuracy I should say 'the *more* concrete descriptions', since a given description can be concrete relative to one set of concepts and abstract relative to another. There is, however, one sense in which we might take a quantitative concept to be concrete *simpliciter*: if there are no concepts relative to which it is abstract except more precise specifications of the quantity itself. For example, *72 miles per hour* is a more precise specification of *velocity over the national speed limit*; and it may only be descriptions like this that provide more concrete correlates of *velocity over the national speed limit*.

4 Other descriptions may be necessary, just not ones in the same set. Washing dishes, for instance, may or may not count as work; in order to do so, an activity requires the correct relation to the concepts of *leisure, preference, wage, boredom* and so on. This is in part how abstraction and supervenience differ. (In the cases I look at here, however, the abstractions do supervene on their more concrete base since they in one sense *reduce* to this base.)

5 Research for this paper was supported by the Modelling in Physics and Economics project at the Centre for Philosophy of Natural and Social Science at the London School of Economics. I am very grateful for this support as well as for all the contributions of the other members of the project.

Bibliography

Bardeen, J., Cooper, L. N. and Schrieffer, J. R. (1957) 'Theory of super-conductivity', *Physical Review* 108.

Cartwright, N. (1989) *Nature's Capacities and their Measurement*, Oxford: Clarendon Press.

Cartwright, N. (1994) 'Fundamentalism vs. the patchwork of laws', *Proceedings of the Aristotelian Society* 103: 279–92.

Cartwright, N. (1999) *The Dappled World: A Study of the Boundaries of Science*, Cambridge: Cambridge University Press.

Lessing, G. E. (1759 [1967]) *Abhandlung ueber die Fabel*, Stuttgart: Philipp Reclam.

Lindsay, R. (1961) *Physical Mechanics*, Princeton: D. Van Nostrand.

Messiah, A. (1961) *Quantum Mechanics*, Amsterdam: Noord-Holland.

Neurath, O. (1933) 'United science and society' in B. McGuiness (ed.) (1987) *Unified Science*, Dordrecht: Reidel.

Index